Landscape Urbanism and
Green Infrastructure

Landscape Urbanism and Green Infrastructure

Special Issue Editor

Thomas Panagopoulos

MDPI • Basel • Beijing • Wuhan • Barcelona • Belgrade

Special Issue Editor
Thomas Panagopoulos
University of Algarve
Portugal

Editorial Office
MDPI
St. Alban-Anlage 66
4052 Basel, Switzerland

This is a reprint of articles from the Special Issue published online in the open access journal *Land* (ISSN 2073-445X) from 2018 to 2019 (available at: https://www.mdpi.com/journal/land/special_issues/greeninfrastructure).

For citation purposes, cite each article independently as indicated on the article page online and as indicated below:

LastName, A.A.; LastName, B.B.; LastName, C.C. Article Title. *Journal Name* **Year**, *Article Number, Page Range.*

ISBN 978-3-03921-369-6 (Pbk)
ISBN 978-3-03921-370-2 (PDF)

Cover image courtesy of Thomas Panagopoulos.

Contents

About the Special Issue Editor vii

Thomas Panagopoulos
Special Issue: Landscape Urbanism and Green Infrastructure
Reprinted from: *Land* **2019**, *8*, 112, doi:10.3390/land8070112 . . . . **1**

Jon Bryan Burley
The Emergence of Landscape Urbanism: A Chronological Criticism Essay
Reprinted from: *Land* **2018**, *7*, 147, doi:10.3390/land7040147 . . . . **5**

Jackie Parker and Greg D. Simpson
Public Green Infrastructure Contributes to City Livability: A Systematic Quantitative Review
Reprinted from: *Land* **2018**, *7*, 161, doi:10.3390/land7040161 . . . . **22**

Catarina de Sousa Silva, Inês Viegas, Thomas Panagopoulos and Simon Bell
Environmental Justice in Accessibility to Green Infrastructure in Two European Cities
Reprinted from: *Land* **2018**, *7*, 134, doi:10.3390/land7040134 . . . . **48**

Minseo Kim, Christoph D. D. Rupprecht and Katsunori Furuya
Residents' Perception of Informal Green Space—A Case Study of Ichikawa City, Japan
Reprinted from: *Land* **2018**, *7*, 102, doi:10.3390/land7030102 . . . . **71**

Zachary Christman, Mahbubur Meenar, Lynn Mandarano and Kyle Hearing
Prioritizing Suitable Locations for Green Stormwater Infrastructure Based on Social Factors
in Philadelphia
Reprinted from: *Land* **2018**, *7*, 145, doi:10.3390/land7040145 . . . . **91**

Giampaolo Zanin, Lucia Bortolini and Maurizio Borin
Assessing Stormwater Nutrient and Heavy Metal Plant Uptake in an Experimental
Bioretention Pond
Reprinted from: *Land* **2018**, *7*, 150, doi:10.3390/land7040150 . . . . **108**

Marie Luise Blau, Frieder Luz and Thomas Panagopoulos
Urban River Recovery Inspired by Nature-Based Solutions and Biophilic Design in
Albufeira, Portugal
Reprinted from: *Land* **2018**, *7*, 141, doi:10.3390/land7040141 . . . . **124**

Jackie Parker and Greg D. Simpson
Visitor Satisfaction with a Public Green Infrastructure and Urban Nature Space in Perth,
Western Australia
Reprinted from: *Land* **2018**, *7*, 159, doi:10.3390/land7040159 . . . . **140**

Thomas Panagopoulos, Stilianos Tampakis, Paraskevi Karanikola,
Aikaterini Karipidou-Kanari and Apostolos Kantartzis
The Usage and Perception of Pedestrian and Cycling Streets on Residents' Well-being in
Kalamaria, Greece
Reprinted from: *Land* **2018**, *7*, 100, doi:10.3390/land7030100 . . . . **157**

About the Special Issue Editor

Thomas Panagopoulos (Dr) is Professor of landscape architecture with specialization in landscape restoration. He received his MSc in renewable natural resources in 1992 and PhD in forestry and natural environment in 1995. He is currently a member of the Research Centre for Tourism, Sustainability and Well-Being and a member of the coordinating body of the PhD program in innovation and land management. He was the department head of landscape architecture and master's degree director at the University of Algarve, Portugal. He is a reviewer and member of the editorial board of several reputed international journals on sustainability and environmental management. He has acted as a principal investigator, co-principal investigator, and investigator in projects with a total of approved funding of over 8 million euros. This is a result of his research strategy that crosses many disciplinary boundaries to create a holistic transdisciplinary approach to science, and his multicultural background in fostering research at an international level. Further, he has vast experience in working in many European and private projects. Currently, he is coordinating the projects: BIODES "Improving life in a changing urban environment through biophilic design"; RESTORE: Rethinking sustainability towards a regenerative economy; and TrailGazerBid Enhancing natural and cultural assets to stimulate economic development. He has also helped many cities to develop their sustainability plan. From 2011 to 2017, he was in the executive board of the UNISCAPE (the Network of Universities for the implementation of the European Landscape Convention).

Editorial

Special Issue: Landscape Urbanism and Green Infrastructure

Thomas Panagopoulos

Research Centre of Tourism, Sustainability and Well-Being, Faculty of Science and Technology, University of Algarve, Gambelas Campus, 8000 Faro, Portugal; tpanago@ualg.pt; Tel.: +351-289800900

Received: 15 July 2019; Accepted: 16 July 2019; Published: 17 July 2019

Abstract: With the notion of landscape urbanism long neglected, interlinkages between ecology and architecture in the built environment are becoming visible. Yet, the diversity in understandings of the interconnections between cities and nature is the starting point for our research interest. This volume contains nine thoroughly refereed contributions concerning a wide range of topics in landscape architecture and urban green infrastructure. While some papers attempt to conceptualize the relation further, others clearly have an empirical focus. Thereby, this special issue provides a rich body of work, and will act as a starting point for further studies on biophilic urbanism and integrative policies, such as the sustainable development goals of the United Nations.

Keywords: built environment; nature-based solutions; sustainable cities; biophilic design; urban planning; landscape architecture; environmental justice; public perception; well-being

1. Introduction

The global population is projected to grow from 7.7 billion in 2019 to almost 10 billion by the middle of the century, with urban areas to absorb all of the future growth [1]. Rapid urban growth presents an important opportunity for economic prosperity, meanwhile, unsustainable, non-resilient urbanization patterns have caused the degradation of ecosystems and their services. Therefore, urbanization presents one of the most urgent challenges of the 21st century to the implementation of an ambitious urban development agenda that seeks to make cities and human settlements inclusive, safe, resilient, and sustainable (according the 11th goal of the United Nations 2030 agenda for sustainable development) [2].

Green infrastructure is a network of green spaces designed and managed to deliver a wide range of ecosystem services that can improve environmental conditions and therefore citizens' health and quality of life [3]. As cities grow bigger, it is imperative to maintain or increase ecosystem services per inhabitant. Restoring, rehabilitating, and increasing connectivity between existing, modified, and new green areas within cities and at the urban–rural interface is necessary to enhance the adaptive capacity of cities to cope with the effects of changes and to enable ecosystems to deliver their services for more livable, healthier, and resilient cities [4].

The underlying economic conditions and the need for urban growth due to the growing population require environmentally sustainable policies in order to address the problem in accordance with a healthy environment. Cities already find themselves in a challenging context facing risks associated with climate change, increasing health crises, social inequality, and global competition [5]. A paradigm shift is needed towards restorative sustainability for new and existing urban areas, and increasing efforts must be made to ensure that multidisciplinary knowledge is adequately taken into consideration. Doing so will help promote solutions that celebrate the richness of design creativity while enhancing users' experience, comfort, health, well-being, and satisfaction, and will allow for improved harmony between urban and natural ecosystems, thus helping to reconnect urban dwellers to nature.

To address these issues, the prime aim of this Special Issue was to provide a set of innovative contributions regarding the links between cities and nature. Furthermore, it focused on the emerging opportunities and challenges of landscape architecture, as innovative nature-based solutions and climate change adaptation issues require transdisciplinary research. This collection of papers provides approaches and methodologies that are useful for both researchers and professionals. It contains nine thoroughly refereed contributions, accepted through a single-blind review process following standard MDPI review guidelines.

The Special Issue consists of the following papers: Jon Bryan Burley [6] conceptualizes the emergence of landscape urbanism in a form of chronological criticism, presenting a broad historical overview, comparing the normative theories derived in the Western traditions embedded in urban design with the general values of landscape urbanism, and revealing the transdisciplinary perception in which the planning and design community derived the foundation of landscape urbanism. Parker and Simpson [7] present a systematic quantitative review on how public green infrastructure contributes to city livability. This review informs urban planners, decision makers, and researchers about the psychological, physiological, general well-being, and wider societal benefits that humans receive as a result of experiencing nature in urbanized landscapes.

De Sousa Silva et al. [8] investigate the issue of environmental justice focusing on availability and accessibility to green infrastructure in two contrasting European cities. Quantitative indicators of public green space revealed inequalities between prosperous city districts and suburbs where minorities live. Urban planners were informed on how to balance green space distribution within city neighborhoods, providing environmental justice without provoking green gentrification. Kim et al. [9] examined the potential of informal green space as supplementary urban green space to meet the well-being needs of residents. They conducted a study on residents' perception in Ichikawa, Japan, a shrinking and aging city that is clearly deficient in urban green areas, currently providing only 3.43 m^2 of green space per capita. Their results revealed that informal green space is recognized by residents and can play an important role in providing green infrastructure services in cities with spatial and financial limitations, thereby relieving the burden of governments and helping them meet the needs of residents. Meanwhile, the elderly and people in lower socioeconomic groups often experience unequal availability of green space. Urban planners should be aware of this environmental justice issue and address this into their green infrastructure policies.

Christman et al. [10] developed a new framework to support decision making regarding green stormwater infrastructure implementation in Philadelphia. They employed a participatory approach using a diverse set of variables that evaluate suitable sites, and integrated social factors in site prioritization based on their ranked proximity to a variety of features defined by the built and social environments. The results of this study indicate optimal locations in the city for the implementation of tree trenches, pervious pavement, rain gardens, and green roofs. Zanin et al. [11] assessed stormwater nutrient and heavy metal plant uptake in a bioretention pond in Italy in order to study a solution based on sustainable urban drainage systems (SUDS). Eleven species of herbaceous perennial helophyte plants, with ornamental features, were used and tested to reduce and treat stormwater runoff in urban areas. Blau et al. [12] demonstrate nature-inspired solutions for the recovery of an urban river of South Europe that was canalized and transformed in culvert pipes. In the face of climate change, the river restoration project presents a unique opportunity for adaptation to its consequences and to provide areas for recreation and contact with nature within the built environment. Using a regenerative sustainability approach based on biophilic design principles, it was proposed to re-naturalize the river corridor that once was crossing the old town of Albufeira in Portugal as a way to improve well-being and city resilience in the long term. Such actions demonstrate the benefits of the transition to a regenerative economy.

Parker and Simpson [13] undertook a study of visitor satisfaction with a public green infrastructure and urban nature space in Perth, Australia, using the importance-performance analysis technique. The survey informed the green managers about the needs for improvement of the amenity and infrastructure, and also optimized nature space management, directing attention towards a more effective utilization of scarce resources. A similar study was also conducted by Panagopoulos et al. [14] in Kalamaria, Greece. They investigated residents' perceptions and satisfaction rates concerning the pedestrian and cycling streets of the city in times of economic crisis, and evaluated their importance for residents' well-being. The survey showed frequent and longtime use of the pedestrian zones. Even that the urban landscape aesthetics and people's health and well-being were considered as important functions of pedestrian zones, at the same time, residents were not satisfied with their quality of life and the existing green infrastructure. The research shows that local authorities can use participatory approaches in re-designing and transforming public spaces and managing a city's green infrastructure, and that the information gained from participatory approaches can be used to increase well-being in cities.

2. Conclusions

In conclusion, these papers unambiguously demonstrate an important contribution from landscape architecture theory combined with in situ observations based on participatory approaches and tools like nature-based solutions and geographic information to promote equitable green infrastructure in a sustainable urban planning framework. The special issue addresses a broad range of different topics, leveraging on the multidisciplinary vision of landscape urbanism. The papers suggest a diversity in understandings about the connection between cities and nature. Innovative urban design and planning may reduce environmental burdens, foster equitable access to public spaces, and promote sustainable urban mobility patterns. Moreover, the implementation of green infrastructure may increase city resilience to climate change and disaster risk reduction. Thereby, this special issue provides evidence on practices and lessons learnt regarding green infrastructure and biophilic urbanism, thus contributing to the sustainable development goals of the United Nations.

Acknowledgments: The editor expresses his gratefulness and gratitude to all reviewers for their support and their critical and constructive comments for these manuscripts. This has improved significantly the quality of this collection. Finally, the editor would like to thank the editorial assistance office of MDPI for their support throughout the review and publication process of this Special Issue. This work was partly financed by the FCT-Foundation for Science and Technology through project PTDC/GES-URB/31928/2017 "Improving life in a changing urban environment through biophilic design".

Conflicts of Interest: The author declares no conflict of interest.

References

1. United Nations, World Urbanization Prospects 2018. Available online: https://population.un.org/wup/Publications/ (accessed on 12 July 2019).
2. United Nations, Sustainable Development Goals. Available online: https://sustainabledevelopment.un.org/post2015/transformingourworld (accessed on 12 July 2019).
3. Panagopoulos, T.; Duque, J.A.G.; Bostenaru Dan, M. Urban planning with respect to environmental quality and human well-being. *Environ. Pollut.* **2016**, *208*, 137–144. [CrossRef] [PubMed]
4. Berte, E.; Panagopoulos, T. Enhancing city resilience to climate change by means of ecosystem services improvement: A SWOT analysis for the city of Faro, Portugal. *Int. J. Urban Sustain. Dev.* **2014**, *6*, 241–253. [CrossRef]
5. Lovell, S.T.; Taylor, J.R. Supplying urban ecosystem services through multifunctional green infrastructure in the United States. *Landsc. Ecol.* **2013**, *28*, 1447–1463.
6. Burley, J.B. The Emergence of landscape urbanism: A chronological criticism essay. *Land* **2018**, *7*, 147. [CrossRef]
7. Parker, J.; Simpson, G.D. Public green infrastructure contributes to city livability: A systematic quantitative review. *Land* **2018**, *7*, 161. [CrossRef]
8. De Sousa Silva, C.; Viegas, I.; Panagopoulos, T.; Bell, S. Environmental justice in accessibility to green infrastructure in two European cities. *Land* **2018**, *7*, 134. [CrossRef]

9. Kim, M.; Rupprecht, C.D.D.; Furuya, K. Residents' perception of informal green space—A case study of Ichikawa City, Japan. *Land* **2018**, *7*, 102. [CrossRef]
10. Christman, Z.; Meenar, M.; Mandarano, L.; Hearing, K. Prioritizing suitable locations for green stormwater infrastructure based on social factors in Philadelphia. *Land* **2018**, *7*, 145. [CrossRef]
11. Zanin, G.; Bortolini, L.; Borin, M. Assessing stormwater nutrient and heavy metal plant uptake in an experimental bioretention pond. *Land* **2018**, *7*, 150. [CrossRef]
12. Blau, M.L.; Luz, F.; Panagopoulos, T. Urban river recovery inspired by nature-based solutions and biophilic design in Albufeira, Portugal. *Land* **2018**, *7*, 141. [CrossRef]
13. Parker, J.; Simpson, G.D. Visitor satisfaction with a public green infrastructure and urban nature space in Perth, Western Australia. *Land* **2018**, *7*, 159. [CrossRef]
14. Panagopoulos, T.; Tampakis, S.; Karanikola, P.; Karipidou-Kanari, A.; Kantartzis, A. The usage and perception of pedestrian and cycling streets on residents' well-being in Kalamaria, Greece. *Land* **2018**, *7*, 100. [CrossRef]

Essay

The Emergence of Landscape Urbanism: A Chronological Criticism Essay

Jon Bryan Burley

School of Planning, Design, and Construction, Michigan State University, East Lansing, MI 48824, USA;
burleyj@msu.edu; Tel.: +1-989-682-4284

Received: 31 October 2018; Accepted: 16 November 2018; Published: 27 November 2018

Abstract: Scholars and practitioners have great interest in topics related to spatial patterns and the organization and properties of space. Landscape urbanism is one of these topics of interest. This essay, in the form of chronological criticism, presents a broad historical overview of the rise of landscape urbanism, primarily from a landscape architectural/geographical/ecological perspective, comparing the normative theories derived in the Western traditions embedded in urban design and architecture with the general values of landscape urbanism. In part, the essay employs the metaphors of Euclidean/Cartesian mathematics and fractal geometry to illustrate these differences. At the conclusion of the article, the reader should understand the historical context in which the planning and design community derived the emergence of landscape urbanism.

Keywords: urban design; landscape first; post-postmodernism; landscape history; urban ecology; plant ecology; context-sensitive design; landscape theory; urban geography

---

1. Introduction

This study explores the differences in perception and thinking about urban space from normative theories affiliated with deeply imbedded traditional urban-design values in comparison with the emerging beliefs associated in the landscape urbanism movement. It is the story of two somewhat divergent perspectives, long dominated by one perspective. Neither perspective is correct or right. Rather, both perspectives have much to offer, as both are normative theories, meaning they are ideas guiding the art of decision-making [1,2]. The arena of normative theories is quite different than the realm of scientific theories and creating predictive models. All normative theories are falsifiable, meaning that there are always examples where normative theory can be demonstrated to be untrue [1]. However, normative theories guide the painter, musician, architect, lawyer, and medical doctor on how to conduct their craft. With all its knowledge, science cannot tell a person what to do, such as what color of paint to put on a canvas, what to say to gain a favorable decision for a client, or where exactly to make an incision on the human body or even whether an incision is in the best interest of the patient—it is all based on judgment [1]. It is unfortunate that science and art have been so separate. Over time, normative theory has been excluded from much of science [1]. Research can rarely advise the artist what to do, but normative theory offers an abundance of advice. The exclusion of normative theory was not always true 100 years ago, when the Doctor of Philosophy degree in the sciences often meant writing essays in criticism and addressing normative theories as well as reporting upon science-based experiments and results [1]. But discussions and exploration of criticism and normative theories have been slowly "weeded-out" of the sciences as scholars adopted the ideas of modern philosophers, with criticism and normative theories remaining primarily in the realm of the arts [1]. Thus, this article is unusual for the times because it is a normative-theory criticism essay in a primarily scientific journal. Yet, no matter how foreign normative theory is to the scientist, normative-theory criticism is essential in the understanding of built form, because built form is governed by many normative theories [1].

Interpretations about built form are heavily based in the perceptions of the past, and shared cultural paradigms of the present. It is not surprising that how one observes and understands space influences how one describes space. This is true for urban environments. To understand how this perception influences decisions, one of the great American plant ecologists, J.T. Curtis [3] (p. 70), notes:

> ... the most important decisions made by an ecologist is that made when he [she] stops his [her] car [4]. In other words, the choice of a place to study is more likely to affect the results than anything the ecologist does subsequently. There is no feasible way whereby this subjective judgment can be completely avoided.

This means that where one looks influences what one will find. In addition, the late anthropologist Lewis Binford (1931–2001) suggests in many words that, in anthropology, more may be learned/ gleaned about the transcriber than the thing/culture itself being described, and the act of unraveling the obtainable knowledge is a difficult task [5]. Binford implies that how one looks at the world affects the description. Burley and Machemer note that, when Western culture discusses other cultures, there is propensity to address architecture of the culture being studied, geomancy/astronomy knowledge and application, and unusual mythical beliefs [1]. Such an approach rarely fairly characterizes the culture, but it does represent what is important to Western biases. Such topics make for profitable books, manuscripts, and interesting stories, but these stories may not be an accurate representation of the character of the culture. Yet Western culture may be completely unaware of this bias and desire to selectively look and describe others. Bateson [6] notes that schoolchildren may recognize this bias even though it is not directly taught in schools, only indirectly by what is and is not discussed, but ironically it does not go unnoticed and is part of the educational process. In some ways, this article addresses the difference in the biases of Western culture related to urban design, landscape, and architecture with what the "schoolchildren" noticed.

Historical Background (from Ionian Greeks to Frank Lloyd Wright, an American)

In the Western world, drivers concerning the perception of space can be observed in the values and knowledge found in Euclidean and Cartesian mathematics and geometry [7]. Space is ordinated into three dimensions with points, lines, areas, and volumes. Boyer describes the quests of many scholars from the Middle East and Africa to Western Europe who attempted to unravel this knowledge [7]. This numerical and spatial knowledge was applied to music, biology, architecture, and art in search for broad and sweeping attempts to integrate the observable world with grand insights and explanations, as illustrated by Doczi [8]. The natural vocabulary for urban design would naturally be these points (landmarks and nodes), lines (edges and paths), areas (districts), and volumes (buildings). The late Kevin Lynch (1918–1984) described the world with much of this vocabulary [9]. One could almost predict that this would be the vocabulary for urban design in the Western world. This vocabulary has been widely adopted and applied in urban and site design, at times without question. In part, great historic planned and designed environments can be explained with concepts related to mathematics and geometry combined with an understanding in astronomy, allegory/metaphor/concept, religion, politics, economics, and technology, such as in Malta [10], Stonehenge in England [11], the Egyptian pyramids at Giza [1], Villa Lante in Italy [12], Bom Jesus do Monte in Portugal [1], Vaux le Vicomte and the Versailles in France [1,13], and Stourhead in England [14]. These governing ideas date back in the literature to at least Vitruvius (translated by the late Morris Hicky Morgan in 1914 and reprinted in 1960 [15]) and exemplified by the early Greek/Roman cities Didyma, Miletos, and Priene, Hierapolis, Ephasus, and Aphrodisias in Anatolia (western Turkey) [16,17]. With the exception of Miletos, these Ionian cities are often overlooked in the history of urban form, yet are formative members of the Euclidean/Cartesian development of built form.

For many modern Greeks, the Ionian setting was the origin of Hellenistic science, technology, and philosophy, with close ties to Athens, where Athenians advanced and developed the early teachings and abilities of the Ionians who fled a conflict with the Persians [1]. Priene had an acropolis, high above the city, and even a temple for Egyptians in the city, plus an early temple to Athena within the city (designed by the architect Pytheos, designer of the Mausoleum at Halicarnassus, one of the seven wonders of the world). Didyma (Didim) was the location of the sacred spring where Zeus and Leto conceived Apollo and Artemis. A series of temples were built and rebuilt near the spring, a relationship of the connection between landscape and built form that was prevalent across Greek culture [18]. The great temple to Apollo was built, constructed/revised even in Roman times. The temple was destroyed in a great earthquake. Miletos (Melitus) is famous for its Hippodamos gridiron city plan, influencing the city plans of many cities around the world. This general planning and design approach has continued for millennia, and is illustrated in recent times by the creation of the classic postmodern urban park, Parc de la Villette in Paris, France, designed by Bernard Tschumi and the beautiful civic areas of Washington, DC; Chicago, Illinois; Camberra, Australia; and New Delhi, India [1,19]. The perception of urban planning and the built environment has been greatly influenced by observing the world through this lens. Since the world was almost exclusively described with this lens, points, lines, areas, and volumes seemed to be a complete and exhaustive set of the potential physical objects that could be possible in an urban setting. It is like believing and understanding the world through the set of whole positive numbers ranging from zero (none) to one, then two, three, to infinity, without the realization that there are negative numbers, fractions, rational, irrational, and imaginary numbers, and many more unusual numbering systems and numerical sets.

The story of the evolving development of urban-landscape theory, planning, and design is similar to the evolving perceptions and advancing knowledge in mathematics, anthropology, geography, and ecology [1]. The difference is that advances in urban design are often lagging behind the sciences and some arts because inventing a new numerical system or new style of painting only takes one person with the conviction to explore the new area, but urban design is a collective societal activity that often requires community consensus and agreement [1].

The evolving understanding about the built environment and space is illustrated through advances at understanding Stonehenge [11]. At one time, the focus was upon the stone objects associated with the henge, but the surrounding landscape is filled with artifacts and structures. Paul Burley, an engineer, and geologists mapped and studied the greater landscape, unveiling insights into the greater design of the area, and suggesting that the landscape was a mirror of a portion of the sky, the same portion of the sky that Egyptians mirrored and revered by cultures in East Africa [1]. The use and organization of the environment is more than the most noticeable remaining objects.

The perspective of some landscape architects, geographers, anthropologists, and ecologists have at times been quite different than the perceptions of those in some engineering and architectural normative principles. This is like the difference between the writings of Vitruvius as opposed the writings of Pliny the Elder (Gaius Plinius Secundus reprinted in 1991 [20]) and Strabo (the complete works translated and published in 2016, a work covering 4804 pages [21]). More recently, it is like the differences between the teachings and understanding of space by the Bauhaus architects and the thoughts and beliefs about design from May Theilgaard Watts and Jens Jensen [22,23]. The Bauhaus artists and architects were indirectly influenced by Viollet-le-Duc (the Purple Duke and the father of modern architecture theory), who wrote about having a concept for a design, the importance of organic design (shapes, forms, relationships, sizes, and adjacencies to organize space), new materials in built form, and the importance of the design process [24]. The Violet-le-Duc, whose extensive writings in French influenced Antoni Gaudi, Frank Furness, Bernard Maybeck, Louis Sullivan, and, in turn, greatly influenced Frank Lloyd Wright. Wright stated, "I thought [Viollet-le-Duc's] Raisonné was the only sensible book on architecture in the world. I later obtained copies for my sons. This book alone enabled us to keep our faith in architecture, in spite of architects." [24] (p. back cover). For the Bauhaus academics, however, the landscape was too complex for such thoughtful development (compare the bland Mies van der Rohe Federal Plaza in Chicago with the complexities of Millennium Park designed by landscape architects and the ecological community in the greater Grant Park setting, along with the inclusion of great, notable architecture and art) [1]. The drivers for this different type of thinking are broadly ecological, understanding many connections from setting to setting, treating each location as unique (culture, economics, function, ecology, and aesthetics), forming larger collective patterns, and associations into a somewhat global system of seemingly infinite diversity. The dividing lines between humans and nature is complex, messy, and not fully understood.

Frank Lloyd Wright seemed to intuitively understand this emerging perspective, and had much more insight about the integration of structure and landscape, traveling to Japan and experiencing the Chinese-inspired designs of the Japanese; plus, he was in contact with Jens Jensen in Chicago, and he was able to educate himself. Wright was able to blend landscape and architecture in a more holistic manner, an approach more closely tied to Chinese design philosophies, where structures and landscape blended together to form a residence/home [1]. Wright embraced some of this Asian perspective. Even his design for Broad Acre City embellished this fusion of rural, forested, and agricultural landscapes with the urban form of industry, commercial areas, and residential settings, a very different approach than the design of many Western-based cities [25]. Wright had believed the Broad Acre City idea could be implemented for Greater Detroit, Michigan, but the idea was never realized until recently, when the city of Detroit has been experiencing spatial reorganization [26]. Broad Acre City could be considered an example of landscape urbanism at the dawn of when such ideas were being considered as much more advanced in design development that the garden city visions of Ebenezer Howard nearly 60 years earlier [1]. Wright spent time understanding nature and those who studied nature, responding to the natural environment in a manner quite different from many [27]. The landscape was not a mystery to Wright. He could be thoughtful.

For much of the history of urban form in Western culture, urban design was dominated by the Euclidean/Cartesian mindset, concerned with objects such as buildings, monuments, roads, and civil engineering (ports, bridges, water supply, energy supply, fortification, industry, and commerce). In contrast, a different approach emerged, grew, and developed, something that some now call landscape urbanism. This movement emerged at a general time when landscape ecology and ideas about sustainability had also emerged. In many ways, it is a different way of thinking about the environment. This paper addresses the rise of this alternative manner of considering the urban setting and managing both the built and natural environment. The intent is not to identify which is better or best, but to support understanding and explain the development of these ideas. This essay is also not a detailed explanation of landscape urbanism, its nuances, and current debates; instead, the essay explains the origins and rise of landscape urbanism in a broad sense.

2. Methodology

The methodology employed in this essay is a method of scholarship that is common to the design arts including architecture and landscape architecture, but it is a method of inquiry that has slowly disappeared from much scientific inquiry. The method is termed "criticism". This methodology is meant as a means to address and assess normative theory. French society has embraced the art of criticism as a national pastime, with several widely read competing publications ranging from well-written essays addressing political, performing-arts, and fine-arts topics, advancing the understanding and appreciation of new knowledge to the art of lampooning and satire, illustrated by Critique (ISSN 0011-1600), Les Temps Modernes (ISSN 0040-3075), and Charlie Hebdo (ISSN 1240-0068).

Criticism is a broad term and does not necessarily mean finding fault with an idea or project. Often it consists of bringing clarity, understanding, and comprehension to new, emerging, and avant-garde ideas. In the field of ecology, the Ecological Society of America has a scholarly forum, Frontiers in Ecology and the Environment, where it allows the membership to offer criticism concerning normative theories affiliated with ecological developments. For example, the 2017 issues feature translational ecology (linking ecological science with the normative theory world of decision making) with articles by Wall, McNie, and Garfin and Safford (et al.) [28,29]. In the field of history, criticism is a common activity, as scholars bring insight, explaining the events of humanity, writing books concerning these events, exemplified by Cranz and Gothien [30,31]. These books are considered to be quality academic achievements. In philosophy, manifestos are written explaining normative new theories concerning the meaning of human existence, as illustrated by Foucault and Faubion [32,33]. Therefore, criticism is found in numerous academic fields as a common form in inquiry. At times, criticism has been lacking in planning and design (see Appendix A).

The format for writing criticism is somewhat open and flexible. It is a narrative. The investigator progressively leads the reader through a series of connected comments and perspectives arriving at a concluding statement(s). Often credentials and life-long experiences of the author(s) add credibility to the statements made in the criticism. Typically, the perspective of senior academics and practitioners are valued for their critical thoughts.

This essay addresses the evolution in thought concerning landscape urbanism with connections to sustainability and landscape ecology, employing a somewhat chronological format (see Appendix A). The essay selects key formative moments (in the opinion of the author) in the development of these ideas, arriving at the present. The essay winds, at times, throughout history as the author selects examples to illustrate ideas (see Appendix B).

3. Discussion/Results

There are moments in the development of landscape and urban-design history where one might begin describing the development and current thinking concerning the evolution of landscape urbanism and related concepts. For this discussion, the choice is Birkenhead Park across the Mersey River from Liverpool, England (Figure 1). The park was designed by Joseph Paxton, a blend of landscape designer and architect [1]. This is the park that inspired Frederick Law Olmsted Sr. in the design of Central Park, New York; however, Central Park is clearly delineated between the softscapes of the park and the hardscape of architecture across the street surrounding the park. At Birkenhead Park, the site is a mixture of central softscapes with a periphery of buildings within the site and a circulation system for pedestrians and for carriages. In other words, the boundaries between hardscape and softscape are integrated. This difference often goes unnoticed as authorities typically explore the similarities in the circulation system and softscape features between the two parks.

Figure 1. View of Birkenhead Park in 2015, a mixture of open space with a variety of active and passive uses, and periphery housing and structures within the park (copyright © 2005 Jon Bryan Burley, all rights reserved and used with permission).

For many observers, the classification of space and land creates distinct districts such as at Central Park—urban and parkland—but Birkenhead Park is an attempt to integrate the two. While the term landscape architect can be attributed to an adaptation of the French term "architecte paysagiste"; the English term "landscape" architect was first used by Meason in 1828 to describe the integration of architecture and landscape in a picturesque fashion, suitable for painting. Birkenhead Park is a less picturesque but still beautiful attempt to integrate landscape and multiple structures [34]. This type of integration was not new as illustrated in Burley and Machemer [1]; now, however, urban form was being explored in new ways.

The differences between traditional urban design and the evolving landscape-urbanism movement has a parallel in plant ecology. Real, L.A., and J.H. Brown (1991) present papers by both eminent ecologist Frederic Clements (1874–1945) and the great taxonomist H.A. Gleason (1882–1975) [35]. Clements believed that vegetation communities were discrete clustered units, while Gleason suggested that every stand was unique. At the time, conventional wisdom was with Clements; Gleason withdrew back to taxonomy. However, Curtis, McIntosh, and Whittaker provided evidence that Gleason was indeed correct [3,36,37]. The controversy between Clements and Gleason illustrates how science evolves, sometimes bumpy, uneven, and uncomfortable with passionate advocates. The same can be said concerning ideas in planning and design.

In the past, urban designers and some architects interpreted the urban environment in a manner similar to Clements plant-ecology vision (dividing the environment into discrete groups), and some landscape architects interpreted the environment in a manner similar to Gleason, ecologists, and some geographers (seeing the environment less distinctly and more as a continuum from wilderness to the dominating urban architecture and built forms). This does not mean that either is correct or wrong, but rather that the normative values that each broad general group had then drove their perception and organization of space.

Shortly after the creation of Birkenhead Park, some American urban environments such as in Boston/Brookline, Massachusetts, and Minneapolis, Minnesota saw the development of urban green infrastructure in the form of connected greenways [1,38–40]. Individuals such as Frederick Law Olmstead Sr. (1822–1903) and Horace W.S. Cleveland (1914–1900) were concerned with sustainability and the environment long before it became fashionable in current culture. The Emerald Necklace in Boston is a series of connected greenways that began in the late 1870s, designed by Olmsted (Figure 2). The Minneapolis park system contains a complete, citywide, greenway circular park system following lakes, wetlands, creeks, rivers, and boulevards. The system was designed by Cleveland and expanded by others. Every trained landscape architect knows this history and early design precedents because in their training, landscape architects must take a course in the history of their profession, something that is missing in the training of many environmental researchers in many fields at universities, unaware of spatial precedents. Sustainability in the urban environment in the Americas has a tradition that is over 120 years old—surprising to those who believe it is something new.

Another important pre-twentieth century project was Biltmore, a massive mansion and estate constructed on the edge of Ashville, Tennessee from 1889 to 1895 [1]. The mansion was built in a French Chateau-esque style with 250 rooms and 35 bedrooms, 43 bathrooms, and 65 fireplaces, and is an example of Beaux Arts architecture. However, it is in the landscape where thoughtful planning and design occurred. Olmsted developed a bass pond as an erosion-control sediment pond to catch the extensive soil erosion occurring during site development. Scientific forestry coalesced as a profession through the estate's lands. The various land uses on the estate were assigned by examining the suitability of various portions of the landscape to support various agricultural functions. This was at a time when George Perkins Marsh had written Man and Nature (1864), describing the extensive erosion from the deforestation of hills in Turkey and the filling of ports with sediments [1,41]. Responses to develop thoughtful land management and sustainability in the Americas was much earlier than many people in society currently believe. Another feature of Biltmore is the ability of Olmsted to blend the precedents of the French, Italian, and English Landscape Schools, and the American rugged landscape, together [1]. It reflects how some landscape architects think about design across broad areas, where various styles (normative theories) can be integrated (Figure 3). Like Biltmore (an environment larger in area than many cites), a city is a composition of many things: local vernacular, city beautiful/classic, modern, postmodern, functional and industrial engineering, lost space, urban decay, and natural remnants and features. These land uses in the city require organization just as Biltmore was thoughtfully organized.

Figure 2. The Emerald Necklace looking toward Brookline, Massachusetts, an example of a greenway, green infrastructure, and the practice of sustainability (copyright © 2004 Jon Bryan Burley, all rights reserved and used with permission).

Figure 3. A view from the French Allée at Biltmore, with the Italian gardens to the left of the trees, the English landscape school to the left of the Italian gardens, and the naturalistic Bass Ponds even farther left leading to a river (copyright © 2018 Xiaoying Li, all rights reserved and used with permission).

After this era, advances in the art of planning and design were at a standstill (so much had been accomplished from the 1850s until the 1900s), with many landscape architects designing estates following Beaux Arts norms. Even in 1946, with the exception of a few modernist landscape architects such as James Rose, Roberto Burley Marx, Thomas Church, Dan Kiley and Garrett Eckbo [1], the typical curriculum included standard topics illustrated by Halligan in 1946, focusing upon site design that did not deviate much from the concepts of Andrew Jackson Downing, 100 years earlier [42].

However, in subdivision design, changes were emerging from the early experiments of Olmsted at Riverside, Illinois, and Alermin Hitchkoss (1816–1903) at Lake Forest, Illinois [1]. These subdivisions followed the precedent similar to cemetery design started at Père Lachaise Cemetery in Paris France and continued in the Americas with Mount Auburn Cemetery in Massachusetts, containing characteristics reminiscent of Paxton's Birkenhead Park with winding carriageways fitting into the landscape [1]. Suburban design advanced with the development of such projects as Radburn, New Jersey by Clarence Stein (1882–1975) and Henry Wright (1878–1936) with cluster housing, redefined public space, and safe walkable pedestrian ways to schools and shopping areas, and Greenbelt, Maryland by Hale Walker and Harold Bursley with its affordable housing and green open space [1]. Yet, after World War II, many communities were planned and designed without regard to these more sustainable and safe design precedents.

While advances in site design were somewhat at a standstill, there was going to be a formative change in the sciences from the 1930s to the 1960s [1]. The works of Aldo Leopold (1887–1948) at the University of Wisconsin developing the science of Wildlife Management and publishing Game Management in 1933 exemplify this change [43]. Investigations in natural history, environmental science, agriculture, social science, building technology, and engineering began to be substantial. By the 1960s, individuals such as Ian McHarg could rely upon the knowledge published in Ecological Monographs to assess the landscape for building suitability of barrier islands along the Atlantic coast [44]. Knowledge about the effects of Dichlorodiphenyltrichloroethane (DDT) upon American robins (*Turdus migratorius* L. 1766) at Michigan State University aided Rachel Carson in her writing of Silent Spring [45]. Science had evidence to guide/shepherd the normative theories on planning and design. The combination of modernism, where each design is unique, responding to the qualities and needs of site, client, and users, all driven by the design concept to give form to solutions with the environmental/ecological movement, gave rise to new solutions in both architecture and landscape architecture [1,46].

An important event in the classification of urban space was conducted by Brady et al., but seems to have often gone unnoticed [47]. They studied the urban environment in a manner similar to how Curtis and Whittaker studied woodlands. They could classify suburban environments (buildings less than three stories tall) as urban savanna. Areas with tall buildings were classified as cliff detritus. They discovered that the urban savanna was one of the most productive and diverse environments in their local region [48]. The Brady et al. classification system is very useful when studying built environments across the urban landscape [47]. It represents a new way of thinking about the built environment. Instead of classifying the built environment in terms of its buildings, zoning boundaries, and taxable zones, the environment was classified as an ecological entity that included the activities of humans. There now was a way to blend the continuum of naturalistic landscape settings with the urban landscape in an ecological manner and represent this continuum with maps. Thirty years later, Ellis et al., in 2013, produced a map of Earth representing a version of this continuum, naming the map series the Anthropogenic Biomes of the World [49]. The maps blends the environmental character of both nature and humans.

Mapping and classifying space has often been a challenge. Maps often contain lines and divisions, creating regions and subsets. These maps were often created with the impressions and experience of senior academics and scholars who were widely traveled, observing many conditions and settings. Often these maps are accepted without much controversy—after all, they were all created by authorities on the topic. However, methods in statistical geographical data analysis facilitated understanding the composition and clustering of homogeneous spatial units into groupings on maps, and removed some of the subjectivity associated with classification.

An article by Qi et al. in 2012 illustrates this problem in classifying the environment. When humans draw maps of regions, they often have preconceptions in drawing the maps [50]. Regions are often of similar size, and each region is typically continuous. Yet the maps composed with data ordinated by the computer may be considered to contain fewer biases. Qi et al. discovered that, when studying the spatial distribution of trees in Michigan, there were discontinuous homogeneous regions, large patches of regions within regions, and the size of these regions greatly varied [50]. In addition, some of the data suggested that Michigan was actually part of just one region dominated by the distribution *Ulmus americana* (American elm), a tree of limited economic importance, often considered a weed tree, and of limited botanical interest when compared to other more interesting trees. But the computer has no bias against American elm, and the data suggested the distribution of this tree could classify the whole state into one set. When Qi et al. employed the significant dimensions of the tree-distribution data into sets and groupings, a highly complex map of the regions of Michigan was produced [50]. This map was not similar to any other map produced by experts. Each map varied in many significant ways. So which map over the various attempts to classify the regions is correct? The situation remains unresolved and illustrates the problem when constructing maps and classifications of a landscape. The data from Qi et al. even suggests that each space is somewhat unique, flowing along several dimensional continuums [50]. A concept that is not that different than the findings of Curtis (1959) and Whittaker (1975), where each stand and setting is unique [36,37].

The emergence of landscape ecology in the 1980s, and the maps generated to study the environment, prompted much enthusiasm. This history may be the most well-known literature by the readership of this journal and so it is not presented in this essay. One of the premises of landscape ecology is that the shapes and patterns of homogeneous environmental units influence the use and suitability of space. There was a belief that certain shapes and patterns are more beneficial than other shapes and patterns. However, as these shapes and patterns were examined, it was discovered that one shape and pattern, advantageous to one group of organisms, was equally disadvantageous to another group of organisms. There is no such thing as a universally beneficial pattern. In addition, it is important to note that landscape ecology (studying shapes and patterns) is not the study of the ecology of landscapes, something that is confused and misrepresented by many who study natural history. Indeed, the study of the ecology of landscapes and landscape ecology are inter-related, but geographers and many who study landscape ecology understand this difference and distinction. The field of landscape ecology continues to evolve and grow as investigators explore these shapes, patterns, and relationships.

The problem concerning discovering and devising planned, built, and managed landscape patterns can be explained by examining the habitat-suitability models developed by the United States Fish and Wildlife Service (USFWS). These models were developed by experts to predict species-specific habitat suitability. Upon examination of these models, each wildlife type required a different combination of spatial contents and, at times, different spatial patterns. In other words, each species required a different set of environmental conditions. This implies that a wide variety of patterns and combinations, almost infinite in expression, are necessary to support all organisms on the planet. Early attempts to explore planning and design applying these models, occurred in the late 1980s [51,52]. By 2003, Burley, on-site in Colorado, illustrated that, at best, these models may be able to explain only about one-third of the variance in predicting multispecies preferences for environments [53]. The result means that the universal (best) pattern is a set of nonuniversal patterns, and there is still much that can be learned and discovered in the natural world. This problem concerning optimal design can seem quite contrary to planners and designers who search for solutions to create the best possible spatial conditions. But this mindset to find the best spatial organization originates with designers who are working with sites. Landscape ecologists, geographers, and landscape architects often explore settings and solutions across sites, where the optimal known solution for one site is different from another site.

Ideas about forms, patterns, and designs were going beyond the Euclidean/Cartesian perspective. Fractal geometry was invented several hundred years ago in Italy, but gained momentum in the 1980s, as illustrated by Barnsley and Gleick [54,55]. This approach revealed that spatial values between the whole numbers of 0, 1, 2, and 3, were possible, and the range of values between these whole numbers was infinitely large. Fleurant et al. in 2009 demonstrated how the fractal numbers of individual tree species could be replicated with the box-counting method to generate planting designs for reforestation on large surface mine, plus replicate patterns of lakes and hilltops on the reclaimed landscape [56]. Yue and Burley were able to replicate the general patterns of a traditional Chinese garden in site design by applying fractal studies [57]. Wei, Fleurant, and Burley were able to replicate the patter of buildings on an island in Hong Kong by applying a similar box-counting method [58]. No longer was the framework of Lynch (1960) the only framework possible. Alberti (2009:268) recognized the limitations of Lynch's framework and has called for new and fresh interpretations [59] (p. 268). Applying fractal technology was one of these new interpretations. Many of these fractal objects do not even have names for the different sets and types of fractals. It is easier to say what they are not, or what they are between, than actually define them. In the future, someone may redefine the image of the city with a fractal vocabulary or a new mathematical paradigm.

The combination of a relatively long history in sustainability and green infrastructures, and the ability and skill to read the environment to create new combinations and patterns in built form poised landscape architectures to develop a landscape-urbanism perspective. There was only one thing missing: the experience and ability to plan extensive urban environments. In the 1920s, architects and landscape architects created the urban-planning profession to address the planning of the urban environment [1]. Landscape architects worked with large landscapes, such as national parks and forests, but not necessarily the complete urban environment to form a city. Instead they would work on the design of pieces (sites) or parts of networks. Warren Manning, a landscape architect, worked on a national plan for the nation, but such work was the exception. After World War II, the profession of urban planning drifted towards social and economic planning [1]. Many urban planners could not draw, or design like landscape architects could. Some firms in the United States started employing landscape architects to conduct landscape-planning studies. For example, Eckbo, Dean, Austin, and Williams (EDAW—the "E" in EDAW is for the late Garrett Eckbo, a landscape architect) were hired to produce a comprehensive land-use plan for the whole state of Hawaii [1]. The landscape architects filled the void that was left by American urban planners. Instead of just being a primarily design profession, landscape architecture became a planning and design profession. It was a natural fit but unforeseen 40 years earlier. This was the final skill and ability necessary that led to the rise of landscape urbanism by primarily the landscape architectural community.

Landscape architects explored a different approach in urban form, an approach known as "Landscape First" (as opposed to architecture/circulation first (Figure 4) [1]. In architecture/circulation first, streets are organized, then buildings are placed in the spaces between the building, and green space is then set around a structure as an ornamental feature. In landscape first, the needs of the environment are initially considered—similar in thinking to McHarg in 1969 and Lewis in 1996, so the idea is not necessarily new [44,60]. Circulation is composed to accommodate the needs of the environment. Finally, structures are placed in this setting. It is a very different way of thinking about design and form. Many landscape architects are trained to think this way, and this kind of thinking is illustrated in the works of James Hawks Jr. [61]. In land development, James Hawks Jr. would be the first to explore its possibilities for a client, organizing the landscape to perform ecological services, develop a circulation pattern, and then placing potential structures. Architects and engineers would be involved in the project, but Jim was involved from the start to the end of the project. Dr. Binyi Liu, a professor at Tongji University in Shanghai, has been influential in designing a truly landscape-first project, Jiyang Eco-Park, and has lectured on the subject [1]. Sometimes, this approach has been affiliated with post-postmodernism and landscape urbanism.

Figure 4. The Qing Ting (Dragon Flies) team's design applying landscape-first ideas for Chongming Island, Shanghai, P.R. of China (Copyright © 2014 Jason Simms, Marina Kato, Wu Chengyi, Ya Dan, Luo, Han Liu, Hanxiao Jiang, Yiling Chen, used with permission, all rights reserved).

Critics of the landscape-urbanism movement have noted that some projects in places like New York City are only site projects or connective linear systems, but it takes time to completely influence and affect a previously developed urban area. Jiyang Eco-Park is an example of a more complete landscape-urbanism-like project.

To create a landscape-first environment implies that one has an understanding of the natural world in relationship to the built form. This is not an easy task. There are numerous good readings on urban ecology, such as McDonnell, Hahs, and Bruste; Marzluff, Shulenberger, Endlicher, Alberti, Bradley, Ryan, and Simon; and Berry and Kasarda [62–64].

However, reading such informative material, one could still be perplexed about what to do. How to apply this information? The problem is that knowing is not the same as being able to do. Contributions such as Ecological Urbanism attempt to bridge this knowledge with essays and project examples [65]. Still, understanding how to apply ecological knowledge is often not easy.

Coinciding with the landscape-first movement, post-postmodernism/context-sensitive design was emerging, as described by Turner (1995) and promoted by transportation landscape architects [1,66]. It is a movement that considers culture, function, ecology, economics, and aesthetics in unison. Good projects consider all five issues together [1]. In contrast, modernism and postmodernism primarily explored the program items' function and aesthetics with different and varying normative theories. Modernism permeated all of the professions. Postmodernism was most expressive in the fine and performing arts, and architecture. The music and films of the French late Serge Gainsbourg (1928–1991) exemplified the postmodern movement [1]. In many respects, the landscape-first and post-postmodernism/context-sensitive design movements are normative theories with similar overlapping thinking.

It was this intellectual setting that led to the landscape-urbanism perspective being first used as a term in Australia and quickly found members who agreed with this general approach. The landscape architectural community could find much to agree with this perspective. This perspective looked at the urban setting as a series components and compositions similar to how one thinks about the contents of naturalistic settings, composed of history, objects, organisms, functions, relationships, often without firm divisions, lines, and edges. The contents can be mixed and combined in endless configurations, and flow from one unique setting to another, just as in the natural world. Like fractal geometry, landscape urbanism transcends scale. This is not to say that landscape urbanism must be designed with fractal geometry, but rather landscape urbanism contains fractal-geometry components. It is more than just the whole-numbers perspective of Lynch.

Classifying the works of designers, and understanding terms and variations in normative theory can be a messy business and it has arrived to the landscape-urbanism community. What it is and what it is not has generated much debate, as has the vocabulary of terms. Newton, back in 1951, brought some common sense to the debate [67]. He said:

> Frankly, I believe it most unfortunate, in a way, that we should ever have found the term 'modern' at all necessary in application to design. One day not long ago a well-known architect was asked by a well-meaning dowager, 'Oh, Mr. So-and-so, you're a modern architect, aren't you?' To which the architect replied, 'Madam, if I were a surgeon, would you ask me if I am a modern surgeon?

He goes on to further state: "Whoever wishes to reach new conclusions would do well to examine and revise the premises, the assumptions, the foundations upon which he has built his understanding, for the chances are that what he needs, more than anything else, is a whole-hearted reorientation of his basic attitude." When one reads this book, one almost gets the impression he could be discussing landscape first, post postmodernism, and landscape urbanism. Newton (1951:2–3) states:

> I would call it the notion of togetherness. According to it, the events of our world, in whatever form they appear to us, are to be seen as inter-related, interconnected always in some way to some degree. The degree may be great or it may be small, but nonetheless it suggests that we would wisely be ever alert to seeing things not as isolated separates but as different aspects of some joint or common phenomenon.

Such a statement applies to landscape urbanism. Landscape urbanism is still in its formative stages. It takes time to build cities that are expressive of landscape-urbanism ideals. It took time to build cities that have city-beautiful components, and these cities are still under development (such as

Chicago and Washington, DC), adding features and civil structures. Finally, there is no guarantee that this movement, emerging primarily from the landscape architectural community, will make any significant contributions to truly improve the broad urban setting beyond a few interesting projects. It is too early to tell.

4. Conclusions

Like any idea, landscape urbanism is expressed in many forms and normative theories. Not everyone agrees. There has been an explosion of literature on the topic, just as there was at one time for modernism and for the short-lived postmodern movement. Even today, there are still many modernists and some postmodernists. But now, there are post-postmodernists, and some of them are embracing the ideas of landscape urbanism, or they may be one and the same. The story of this movement is just beginning. It remains to be seen if landscape planners/designs and the intellectual perspective about the environment emerging from ecologists can create a better urban setting.

This essay explains through the eyes of an ecologist/landscape architect how humankind arrived at this moment. It is unclear what will happen next and how long this movement will last. However, in the forthcoming years and decades, there may be numerous schools of thought on the subject, and numerous built projects expressing the normative theories associated with landscape urbanism, and how it relates to movements in sustainability and the discipline of landscape ecology.

Funding: This research received no external funding.

Conflicts of Interest: The author declares no conflicts of interest.

Appendix A

A strong body of the criticism literature in landscape architectural planning and design is missing within the profession. Bruce Sharky, a professor at the Louisiana State University, produced a series of volumes concerning landscape criticism of built works [68–71]. The volumes are excellent examples of landscape criticism.

Chronological criticism is a form of historical criticism often employed in antiquity studies, such as interpreting the Bible. In historical criticism, the investigator examines the origins and interpretations of ancient texts for intended meaning and historical context. This approach to criticism has been around for about four centuries. It can be used in interpreting the origins of ideas and their evolving development [1]. The well-known works of Bronowski and Burke are popular forms of chronological criticism [72,73].

Chronological criticism is the general form of criticism employed in this essay. The essay examines a time period from early built form to the ideas of the present concerning landscape urbanism.

It should be noted that the historical context of much of what has been developed and evolved is missing in the education of many professions. New ideas to one branch of study is an old idea to another. The focus of much current education concerns immediate practical information with no wisdom in the understanding of context. Supposedly the literature review is supposed to compensate for this immediacy, yet it often focuses upon recent advances without recognition of historical precedence. Often, there is a rush to claim new knowledge. At times, historical understanding is considered unimportant. Such a condition exists for some with the topic of landscape urbanism. Chronological criticism can provide a venue to explain its development and context.

Appendix B

The essay is from the view of a landscape architect (a member of the American Society of Landscape Architects (ASLA) for 36 years and a fellow in this society (for his research accomplishments)), who is also an ecologist (a member of the Ecological Society of America (ESA) for 36 years, a life member, and has been at times a certified ecologist) and a registered landscape architect for 36 years from the United States of America, earning 15 ASLA (state/national)/American

Institute of Architects) (state) awards for writing, designed projects, and research. The author was the 2005 American Society for Mining and Reclamation researcher of the year—not an easy task when competing with agronomists, foresters, engineers, soil scientists, and hydrologists for such a distinction as a planner and designer. The author has had over 400 articles, book chapters, and abstracts published nearing the end of his career. He recently coauthored a 601-paged book describing the history of landscape architecture/environmental design, having traveled to 48 countries around the world [1]. He has been a Fulbright scholar in Portugal in 2003, and a research scholar funded by the French government in 2011–2012, stationed at Agro-campus Ouest, Angers, France, part of the Rennes University system. The author has been teaching in higher education for nearly 42 years, including landscape history, planning, and design theory, and research methods. These are the basic credentials of the author writing this essay, providing a background of experiences and scholarly achievements to create the criticism.

References

1. Burley, J.B.; Machemer, T. *From Eye to Heart: Exterior Spaces Explored and Explained*; Cognella Academic Publishing: San Diego, CA, USA, 2016; ISBN 978-1-63487-685-8.
2. Lang, J. *Creating Architectural Theory: The Role of the Behavioral Sciences in Environmental Design*; Van Nostrand Reinhold: New York, NY, USA, 1987; ISBN 10: 0442259816.
3. Curtis, J.T. *Vegetation of Wisconsin: An Ordination of Plant Communities*; University of Wisconsin Press: Madison, WI, USA, 1959; ISBN 10: 0299019403.
4. Ashby, E. Statistical ecology. II—A reassessment. *Bot. Rev.* **1948**, *14*, 222–234. [CrossRef]
5. Binford, L. *Constructing Frames of Reference: An Analytical Method for Archaeological Theory Building Using Ethnographic and Environmental Data Sets*; University of California Press: Berkeley, CA, USA, 2001; ISBN 10: 9780520223936.
6. Bateson, G. *Mind and Nature: A Necessary Unity*; Dutton: Boston, MA, USA, 1979; ISBN 10: 0525155902.
7. Boyer, C.B. *A History of Mathematics*; John Wiley and Sons, Inc.: Hoboken, NJ, USA, 1968; ASIN B01JXTI8N4.
8. Doczi, G. *The Power of Limits: Proportional Harmonies in Nature, Art and Architecture*; Shambhala: Boulder, CO, USA, 1981; ISBN 10: 1590302591.
9. Lynch, K. *The Image of the City*; MIT Press: Cambridge, MA, USA, 1960; ISBN 10: 0262620014.
10. Trump, D.H. *Malta: Prehistory and Temples (Malta's Living Heritage)*; Midsea Books: Santa Venera, Malta, 2002; ISBN 10: 9990993947.
11. Burley, P. *Stonehenge as above, So Below: Unveiling the Spirit Path on Salisbury Plain*; New Generation Publishing: London, UK, 2014; ISBN 13: 978-1-910162-77-4.
12. Burley, J.B.; Kopinski, E. Villa Lante: Italy's greatest renaissance garden. *Mich. Landsc.* **2014**, *57*, 27–30.
13. Casault, J.; Burley, J.B. Restored Versailles: A French Garden with a message. *Mich. Landsc.* **2010**, *53*, 37–42.
14. Monsma, J.; Miller, T.; Burley, J.B. The Hidden Meanings and Metaphors of Stourhead: England's Premier Informal Garden. *Mich. Landsc.* **2011**, *54*, 50–57.
15. Morgan, M.H., Translator; *Vitruvius: The Ten Books of Architecture*; Dover Publications, Inc.: Mineola, NY, USA, 1960; ISBN 10: 9780486206455.
16. Tulay, A.S.; Akat, H. *Didyma Miletos Priene*; Touring and Automobile Association of Turkey: Istanbul, Turkey, Unpublished work.
17. Mert. *Pamukkale Hierapolis-Aphrodisias*; Mert Basim Yatin Dağitim Ve Reklamcilik Tic Ltd. ŞTİ: Istanbul, Turkey, 2004.
18. Scully, V., Jr. *The Earth the Temple and the Gods: Greek Sacred Architecture*, revised edition; Yale University Press: New Haven, CT, USA, 1979; ISBN 10: 0300023979.
19. Orlandini, A. *Un Architecte/Une Œuvere: Le Parc de la Villette de Barnard Tschumi*; Somogy Éditions d'Art: Paris, France, 2001.
20. Secundus, G.P. *Natural History: A Selection*, reprint edition; Penguin Classics: London, UK, 1991; ASIN B0092GDQSQ.
21. Strabo of Amaseia. *Delphi Complete Works of Strabo—Geography*; Delphi Classics: East Sussex, UK, 2016; ASIN B01JK0KODS.

22. Watts, M.T. *Reading the Landscape of America: An Adventure in Ecology*; Macmillan: London, UK, 1957; ASIN B0006AUYYG.
23. Grese, R.E. *Jens Jensen: Maker of Natural Parks and Garden*; John Hopkins University Press: Baltimore, MD, USA, 1992; ISBN 10: 0801842875.
24. Hearn, M.F. (Ed.) *The Architectural Theory of Violet-le-Duc: Readings and Commentary*; The MIT Press: Cambrdige, MA, USA, 1990; ISBN 10 0262220377.
25. Jin, Y.; Burley, J.B.; Machemer, P.; Crawford, P.; Xu, H.; Wu, Z.; Loures, L. The Corbusier dream and Frank Lloyd Wright vision: Cliff detritus vs. urban savanna. In *Urban Agglomeration*; Ergen, M., Ed.; Intech: Rijeka, Croatia, 2018; pp. 211–230.
26. Burley, J.B.; Deyoung, G.; Partin, S.; Rokos, J. Reinventing Detroit: Grayfields—New metrics in evaluating urban environments. *Challenges* **2011**, *2*, 45–54. [CrossRef]
27. Aguar, C.; Aguar, B. *Wrightscapes: Frank Lloyd Wright's Landscape Designs*; McGraw-Hill Professional: New York, NY, USA, 2002; ISBN 10: 0071377689.
28. Wall, T.U.; McNie, E.; Garfin, G.M. Use-inspired science: Making science usable by and useful to decision makers. *Front. Ecol. Environ.* **2017**, *15*, 551–559. [CrossRef]
29. Safford, H.D.; Sawyer, S.C.; Kocher, S.D.; Hiers, J.K.; Cross, M. Linking knowledge to action: The role of boundary spanners in translating ecology. *Front. Ecol. Environ.* **2017**, *15*, 560–568. [CrossRef]
30. Cranz, G. *The Politics of Park Design: A History of Urban Parks in America*; The MIT Press: Cambridge, MA, USA, 1982; ISBN 10: 0262530848.
31. Gothien, M.L. *A History of Garden Art*; Wright, W.P., Ed.; Archer-Hind, J.M., Translator; Dent and Sons Limited: London, UK; E.P. Dutton and Co, Ltd.: New York, NY, USA, 1928.
32. Foucault, M. *The Archaeology of Knowledge*; Sheridan, A.M., Translator; Routledge Classics: Abingdon, UK, 2002; ISBN 10: 0415287537.
33. Foucault, M. *Michel Foucault: Aesthetics: Essential Works of Foucault 1954–1984: Volume 2*; Faubion, J.D., Ed.; Penguin Books: Londong, UK, 2000; ISBN 10: 9780140259544.
34. Meason, G.L. *On the Landscape Architecture of the Great Painters of Italy*; RareBooksClub.com: Peterborough, UK, 2012; ISBN 10: 1236450248.
35. Real, L.A.; Brown, J.H. (Eds.) *Foundations of Ecology: Classic Papers with Commentaries*; University of Chicago Press: Chicago, IL, USA, 1991; ISBN 10: 9780226705941.
36. Curtis, J.T.; McIntosh, R.P. An upland forest continuum in the prairie forest border region of Wisconsin. *Ecology* **1951**, *32*, 476–496. [CrossRef]
37. Whittaker, R.H. *Communities and Ecosystems*, 2nd ed.; Macmillan: London, UK, 1975; ISBN 10: 0024273901.
38. Pregill, P.; Volkman, N. *Landscapes in History: Design and Planning in the Eastern and Western Tradition*, 2nd ed.; John Wiley and Sons, Inc.: Hoboken, NJ, USA, 1999; ISBN 10: 0471293288.
39. Newton, N.T. *Design on the Land: The Development of Landscape Architecture*; The Belknap Press: Cambridge, MA, USA, 1974; ISBN 10: 0674198700.
40. Tobey, G.B., Jr. *A History of Landscape Architecture: The Relationship of People to Environment*; American Elsevier Publishing Company, Inc.: New York, NY, USA, 1973; ISBN 10: 044400131X.
41. Marsh, G.P. *Man and Nature: Physical Geography as Modified by Human Action*; C. Scribner: New York, NY, USA, 1864.
42. Halligan, C.P. *First Principles of Landscape Architecture*; Edwards Brothers, Inc.: Ann Arbor, MI, USA, 1946.
43. Leopold, A. *Game Management*; Scribner's: New York, NY, USA, 1933.
44. McHarg, I.L. *Design with Nature*; Doubleday/Natural History Press: Garden City, NY, USA, 1969; ISBN 10: 0385055099.
45. Carson, R. *Silent Spring*; Houghton Mifflin: Boston, MA, USA, 1962; ISBN 10: 0395075068.
46. Thayer, R.L. *Grey World Green Heart: Technology, Nature, and the Sustainable Landscape*; Wiley Series in Sustainable Development; Wiley: Hoboken, NJ, USA, 1993; ISBN 10: 047157273X.
47. Brady, R.F.; Terry, T.; Eagles, P.F.; Ohrner, R.; Micak, J.; Veale, B.; Dorney, R.S. A typology for the urban ecosystem and its relationship to large biogeographical landscape units. *Urban Ecol.* **1979**, *4*, 11–28. [CrossRef]
48. Burley, J.; Wang, Y.; Loures, L. New Ecologies: Emergence of the Urban Savanna and Cliff Detritus in a Post-modern Era. In *New Models for Innovative Management and Urban Dynamics*; Panagopoulos, T., Ed.; University of Algarve: Faro, Portugal, 2009; ISBN 978-972-9342-85-4.

49. Ellis, E.C.; Goldewijk, K.K.; Siebert, S.; Lightman, D.; Ramankutty, N. *Anthropogenic Biomes of the World, Version 2, 2000*; NASA Socioeconomic Data and Applications Center (SEDAC): Palisades, NY, USA, 2013.

50. Qi, J.; Wang, S.; Burley, J.B.; Machemer, T. Defining ecological regions in Michigan based on native tree distributions. *Landsc. Arch.* **2012**, *6*, 138–145. (In Chinese)

51. Burley, J.B. Multi-model habitat analysis and design for M.B. Johnson Park in the Red River Valley. *Landsc. Urban Plan.* **1989**, *33*, 261–280. [CrossRef]

52. Burley, J.B.; Johnson, S.; Larson, P.; Pecka, B. Big Stone granite quarry habitat design: HSI reclamation application. In Proceedings of the ASSMR Conference, Pittsburgh, PA, USA, 19–21 April 1988; pp. 161–169.

53. Burley, J.B. Habitat modeling: Spatial landscape assessment at the Rigden Mine, Colorado. In *Working Together for Innovative Reclamation, Proceedings of a Joint Conference the 9th Billings Land Reclamation Symposium and American Society for Mining and Reclamation, and Reclamation 20th Annual National Conference, Billings, MT, USA, 3–6 June 2003*; Barnhisel, R.I., Ed.; ASMR: Lexington, KY, USA, 2003; pp. 103–118.

54. Barnsley, M. *Fractals Everywhere*; Academic Press: Cambridge, MA, USA, 1988; ASIN B00E6T1P00.

55. Gleick, J. *Chaos: Making a New Science*; Viking: New York, NY, USA, 1987; ISBN 10: 0747404135.

56. Fleurant, C.; Burley, J.B.; Loures, L.; Lehmann, W.; McHugh, J. Inverse box-counting method and application: A fractal-based procedure to reclaim a Michigan surface mine. *WSEAS Trans. Environ. Dev.* **2009**, *5*, 76–89.

57. Yue, Z.; Burley, J.B. Non-euclidean methods to characterize the Masters of the Nets Garden, Suzhou, China. In *Recent Researches in Energy, Environment and Landscape Architecture, Proceedings of the 7th IASME/WSEAS International Conference on Energy, Environment, Ecosystems, and Sustainable Development (EEESD '11) and Proceedings of the 4th IASME/WSEAS International Conference on Landscape Architecture (LA '11), Angers, France, 17–19 November 2011*; WSEAS: Madison, WI, USA, 2011; pp. 122–128.

58. Wei, S.; Fleurant, C.; Burley, J.B. Replicating fractal structures with the reverse box counting method—An urban South-east Asian example. In *Peer Reviewed Proceedings of Digital Landscape Architecture 2012 at Anhalt University of Applied Sciences*; Buhmann, E., Ervin, S., Pietsch, M., Eds.; Wichman: Toledo, OH, USA, 2012; pp. 364–370.

59. Alberti, M. *Advances in Urban Ecology: Integrating Humans and Ecological Process in Urban Ecosystems*; Springer: New York, NY, USA, 2008; ISBN 10: 0387755098.

60. Lewis, P. *Tomorrow by Design: A Regional Design Process for Sustainability*; Wiley: Hoboken, NJ, USA, 1996; ISBN 10: 0471109355.

61. Burley, J.B.; Loures, L.; Feng, M.; Hawks, J.W., Jr. ASLA: Polychrome land development in the Upper Midwest. *Int. J. Energy Environ.* **2012**, *6*, 415–423.

62. McDonnell, M.J.; Hahs, A.K.; Bruste, J.H. (Eds.) *Ecology of Cities and Towns: A Comparative Approach*; Cambridge University Press: Cambridge, UK, 2009; ISBN 10: 9780521861120.

63. Marzluff, J.M.; Schulenberger, E.; Endlicher, W.; Alberti, M.; Bradley, G.; Ryan, C.; Simon, U. (Eds.) *Urban Ecology: An International Perspective on the Interaction between Humans and Nature*; Springer: New York, NY, USA, 2008; ISBN 10 0387734112.

64. Berry, B.J.; Kasarda, J.D. *Contemporary Urban Ecology*; Macmillan Publishing Co., Inc.: London, UK, 1977; ISBN 10: 0023090502.

65. Mostafavi, M.; Doherty, G. (Eds.) *Ecological Urbanism*; Lars Müller Publishers: Baden, Switzerland, 2010; ISBN 10: 3037784679.

66. Turner, T. *City as Landscape: A Post Post-Modern View of Design and Planning*; Taylor and Francis: Abingdon, UK, 1995; ISBN 10: 0419204105.

67. Newton, N.T. *An Approach to Design*; Addison-Wesley Press Inc.: Boston, MA, USA, 1951.

68. Sharky, B.; Turner, S.; Womack, W.M. (Eds.) *Critiques of Built Works of Landscape Architecture, Volume 1*; LSU School of Landscape Architecture, Louisiana State University: Baton Rouge, LA, USA, 1994.

69. Sharky, B. (Ed.) *Critiques of Built Works of Landscape Architecture, Volume 4*; LSU School of Landscape Architecture, Louisiana State University: Baton Rouge, LA, USA, 1997.

70. Sharky, B. (Ed.) *Critiques of Built Works of Landscape Architecture, Volume 7*; LSU School of Landscape Architecture, Louisiana State University: Baton Rouge, LA, USA, 2001.

71. Sharky, B. (Ed.) *Critiques of Built Works of Landscape Architecture, Volume 8*; LSU School of Landscape Architecture, Louisiana State University: Baton Rouge, LA, USA, 2003.

72. Bronowski, J. *Ascent of Man*; Little Brown and Company: Boston, MA, USA, 1973; ISBN 10: 0316109304.
73. Burke, J. *Connections*, 1st ed.; Little Brown and Company: Boston, MA, USA, 1978; ISBN 10: 0316115816.

 *land*

Review

Public Green Infrastructure Contributes to City Livability: A Systematic Quantitative Review

Jackie Parker [1] and Greg D. Simpson [2,*]

1   School of Design and Built Environment, Curtin University, Perth WA 6102, Australia;
     j.d.parker@postgrad.curtin.edu.au
2   College of Science, Health, Engineering and Education—Environmental and Conservation Sciences,
     Murdoch University, Perth WA 6150, Australia
*   Correspondence: G.Simpson@Murdoch.edu.au

Received: 21 October 2018; Accepted: 14 December 2018; Published: 18 December 2018

Abstract: Consistent with the *Land Urbanism and Green Infrastructure* theme of this special issue of Land, the primary goal of this review is to provide a plain language overview of recent literature that reports on the psychological, physiological, general well-being, and wider societal benefits that humans receive as a result of experiencing public green infrastructure (PGI) and nature in urbanized landscapes. This enhanced well-being and the wider societal benefits that accrue to urban dwellers as a result of interacting with quality PGI contributes to the concept known as city or urban *livability*. The quantitative analysis and theoretical synthesis reported in this review can inform decision makers, stakeholders, and other PGI and urban nature (UN) researchers of the benefits that urban populations receive from experiencing quality PGI spaces and UN and the contribution those spaces make to the livability of urban areas. With diminishing opportunities for the acquisition of new public open space to increase PGI and re-establish UN near urban centers, the efficient management and continuous improvement of existing PGI and UN is essential to promote and foster opportunities for human-to-nature contact and the known benefits therein derived. In addition to identifying an increased research interest and publication of articles that report on the contribution of PGI spaces to urban livability over the past decade, the review identifies and reports on the seven focus areas of PGI-livability research and the six attributes of PGI spaces that the current literatures report as contributing to the livability of urbanized landscapes. After providing a quantitative analysis for the reporting of those research areas and PGI attributes and summarizing key findings reported in the literature regarding the contribution that PGI spaces make to urban livability, this review also identifies knowledge gaps in the published literature and puts forward recommendations for further research in this rapidly expanding multidisciplinary field of research and policy development.

Keywords: biophilic design; public amenity; public green infrastructure (PGI); public open space; renaturing cities; sustainable development; livability; liveability; urban nature (UN); well-being

1. Introduction

First articulated and popularized by Wilson [1], the Biophilic Hypothesis states that humans have an innate, inbuilt affinity to natural systems and living things. Wilson [1] hypothesized that this is likely to be a by-product of evolution, born instinctively from humanity's heritage of hunter-gatherer focused lifestyles. In more recent times, a growing disconnect between humans and nature (extinction of the nature experience) has emerged [2–4]. This disconnection has significant negative impacts on the general health and well-being of increasingly urbanized human populations [2]. Supporting the reconnection of people with urban nature (UN) is critical to reverse the extinction of nature experience and to access the wide range of physical and mental health benefits provided by quality public green infrastructure (PGI) that incorporates UN [5].

Numerous authors have reported in detail on the contested definitions and inconsistencies in terminology that prevail in this research space (e.g., [6–12]), so before proceeding further it is important to define the terminology as it is applied in this review. Urban green infrastructure (UGI) is suitably articulated by Norton et al. [13] (p. 128) to be a "network of planned and unplanned green spaces, spanning both the public and private realms, and managed as an integrated system to provide a range of benefits. UGI can include remnant vegetation, parks, private gardens, golf courses, street trees and more engineered options, such as green roofs, green walls, biofilters and raingardens". Within this review article, the term PGI is used for consistency and specifically references vegetated public open spaces and urban public green spaces, such as parks and UN spaces [10,13,14]. Urban nature is a UGI element composed of remnant and restored examples of nature indigenous (native) to that locale [14]. Also known as 'indigenous biodiversity', UN spaces should ideally support examples of the micro and macro flora and fauna that would have occupied the area before humans converted the land to an urban matrix. Hereafter, the combination of PGI and UN will be collectively referred to as PGI.

Public green infrastructure affords urban residents with opportunities to exercise, play sports (organized and unorganized), socialize, relax, learn, and experience nature. Aside from the reported psychological and physiological benefits, engaging with PGI has been shown to improve humankind's general outlook on life [15,16]. In recent years, global challenges, such as the compelling evidence that climate change is likely to significantly impact the general health and well-being of human populations [16] and evolving social values [17], have driven research into the contribution that PGI makes to urban livability. Researchers also report that provision of PGI can alleviate several other emerging challenges to urban sustainability through the wider social and environmental benefits provided by such spaces (e.g., [16,18–25]). Specifically, these benefits include enhancing environmental management of underutilized or degraded natural assets, increasing conservation of existing PGI assets, micro-climatic advantages, habitat creation for native wildlife, and/or habitat improvement [18–25].

The concept of city/urban 'livability' emerged during the 1980s, as city planners and theorists attempted to describe and quantify how social, political, economic, and environmental factors contributed to the quality of citizen life in urban settlements (e.g., [26–29]). Giap et al. [7] postulated that livability is a place-based concept that encompasses many factors that contribute to the quality of life and well-being of residents. Giap et al. [7] (pp. 178,179) went on to report that the dependence of livability and the quality of the physical environment of a city on "the performance of key urban systems and processes" had spawned several proxy measures for livability that assigned values to those "systems and processes" based on performance, community perception, and other scale dependent factors. The ratings for those factors are consolidated into a single score to be compared against other cities around the world. Currently, two prominent global urban livability scales are produced annually by the United Kingdom based Economist Intelligence Unit (EIU) and the global Mercer LLC consultancy [30]. Both livability scales feature prominently in media reporting and are heavily utilized in promoting the attractiveness of a city as a place to live and visit, which is a valuable tool for influencing decision making processes of individuals, but both have been criticized for not giving sufficient emphasis to PGI spaces in their metrics [26,31]. Another similar scale is the Monocle Quality of Life Survey [32].

Cities with high-ranking livability scores are sought after destinations. Cities that are considered to be highly livable are perceived to provide social and economic benefits, such as foreign business and housing investments; local and international economic stimulus; increased local community involvement and personal connections; and an increase in individuals sense of pride [27,31]. Tzoulas et al. [33] report that quality PGI spaces play an import role by increasing feelings of attachment to their community among urban dwellers and by providing opportunities for them to interact with other residents.

Public green infrastructure can contribute to urban livability as one of the key urban systems identified by Giap et al. [7] and by providing the social and environmental factors identified by Tzoulas et al. [33]. However, with diminishing opportunities for the acquisition of new public open space to increase PGI and re-establish UN near urban centers, the efficient management and continuous improvement of existing PGI is essential to promote and foster opportunities for human-to-nature contact and the known benefits therein derived. This review provides a quantitative analysis and theoretical synthesis of recent peer-reviewed literature concerning PGI, UN, and the contribution that such spaces make to the livability of urbanized landscapes. Informed by the compilation of the dataset shared via Simpson and Parker [29], we believe this to be the first article to provide a quantitative review of the literature regarding the contribution that PGI makes to urban livability. This systematic review was initially undertaken to inform the design of a questionnaire-based survey that explored the satisfaction of visitors to an urban PGI space [34–36]. The information provided in the Results and Synthesis sections of this review article can, however, inform stakeholders, decision makers, and other researchers regarding the psychological, physiological, and wider societal benefits that urban populations receive from experiencing quality PGI and UN in the context of *landscape urbanism and green infrastructure*, which is the focus of this special issue of *Land*.

2. Methods

2.1. Systematic Quantitative Literature Review

As reported in the Data Descriptor of Simpson and Parker [29], this systematic quantitative literature review is based on the approach of Pickering and Byrne [37] and the Preferred Reporting Items for Systematic Reviews (PRISMA) guidelines (http://prisma-statement.org/) [38].

In December 2016, over 15,000 databases, including Scopus, Web of Science, and all the major English language publishing houses, were searched to identify articles related to PGI, UN, and urban livability using the search terms listed in Table 1. Following the search criteria and PRISMA expression reported in Simpson and Parker [29], two commissioned academic editorial thought pieces, one edited book chapter, and 68 peer-reviewed articles, (hereafter all referred to as 'articles') were identified and deemed suitable for inclusion in this systematic quantitative literature review (Table S1). The PRISMA expression for this systematic review and reported in the Data Descriptor of Simpson and Parker [20] is reproduced here as Figure 1.

As a final measure, a supplementary search using the method detailed above was performed in May 2018 to identify any relevant articles that had become discoverable between January and December 2017. This supplementary search identified another 16 recently published peer-reviewed articles reporting research that was relevant to this review [29]. The 16 articles identified in the supplementary search were analyzed and compared to the 71 articles identified in the initial search.

Table 1. Search terms used to identify papers included in the literature review. Potential papers were filtered using the primary AND secondary search terms.

Primary Search Terms	Secondary Search Terms
"public green infrastructure"	livability/liveability
"public open space"	"city livability/liveability"
POS	"user satisfaction"
"urban open space"	"visitor satisfaction"
"green space"	
"urban nature"	
park	
wetland	

Source: Simpson and Parker [29].

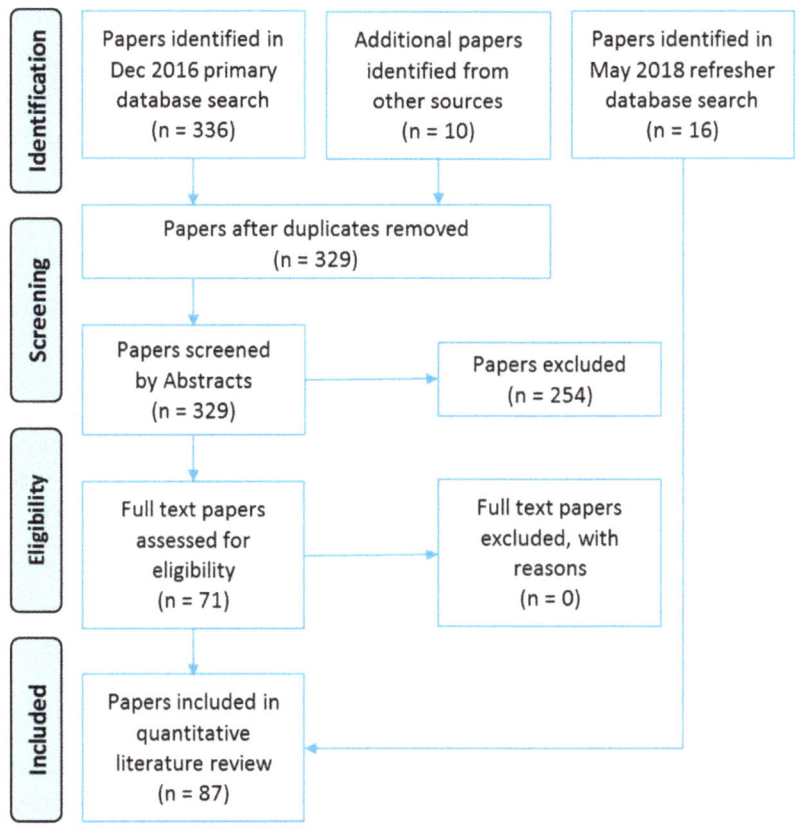

Figure 1. Preferred Reporting Items for Systematic Reviews (PRISMA) Expression for the systematic quantitative literature review. Source: Simpson and Parker [29].

As is standard practice for quantitative reviews, information about the articles are recorded under all/multiple categories for each of the aspects that an article reports against. The full classification of data for each of the articles analyzed under this review is available in the Data Descriptor by Simpson and Parker [29].

2.2. Data Analyses

The data analyses provided in this review utilize both graphical and numerical techniques. Figures that report percentages also incorporate error bars that display the 95% confidence intervals (95% CIs) for the proportions [39]. For a sample size of $n > 10$, if the 95% CI error bars of two categories overlap for half their length or less, then that provides some evidence for a statistically significant difference ($0.01 < p < \approx 0.05$) between those parameters [40]. Similarly, if only the tips of the 95% CI error bars of two categories overlap, then there is evidence for a statistically significant difference ($p < \approx 0.01$) between those parameters [40]. Comparisons of 95% CIs across more than two categories can also be informed by a confirmatory statistical test. In this review, the chi-squared (χ^2) test for goodness of fit was utilized to test the rate of publication in each category against the mean rates of publication across all categories [41] to ascertain if statistically significant patterns could be detected in the rate of reporting with respect to research effort regarding how the attributes of quality PGI spaces contribute to urban livability. Similarly, the chi-squared test for comparison of proportions was utilized to compare between the percentage of articles from the initial search and supplementary search reporting a new tool and also the research focus of articles identified in the initial and supplementary searches [41]. Evidence for patterns in the reporting of factors associated with quality PGI spaces and indicators of urban livability (i.e., measures for human health and well-being, for livability, and for social aspects of PGI) was also investigated using the statistically robust Pearson Coefficient of Correlation. Patterns in the publication of PGI attributes identified as contributing to urban livability were also investigated using the Pearson correlation coefficient. Pearson correlation analyses are insensitive to the effects of nonnormality in datasets with n > 5 and for the type of measurement scale or distribution for n > 30 [42,43]. These characteristics allow Pearson correlation coefficients to be calculated for the binary data presented in Simpson and Parker [29] in order to investigate linkages in the reporting of the factors associated with quality PGI spaces and the indicators of urban livability [44,45]. The significance of all correlations was determined using the similarly robust *t*-test [46,47].

3. Results

3.1. Research Effort

As previously mentioned in the Methods section, the initial search for this review produced a total of 71 articles on PGI and urban livability deemed appropriate for further analysis, from which several trends were identified. The first trend identified was that the rate of publication suggests an increase in the research effort being directed towards understanding the links between PGI and urban livability (Figure 2). This is especially evident with respect to the past six years where a total of 60 articles were published, representing a 122% increase on the 27 articles published in the first 12 years of this century.

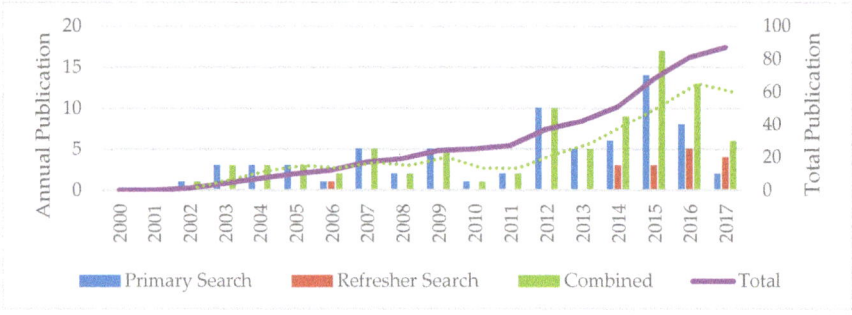

Figure 2. Rate of publication suggests an increase in the research effort being directed towards understanding the links between public green infrastructure (PGI) and urban livability since the year 2000 (n = 71). Green dotted line is the 3-year rolling average for annual publication rate.

3.2. Geographic Distribution of Research

Based on the articles identified in the initial review that reported research in relation to specific cities, with 20 published studies Australia dominates research into the connections between urban livability and PGI, both as a country (Figure 3) and a continent (Figure 4). The majority of those studies focused on the Western Australian state capital of Perth, which is one of the two global hotspots for this type of research (Figure 5). The United Kingdom and the United States of America ranked second and third in terms of countries with reported research into links between urban livability and PGI spaces. Interestingly, the developing nations of Indonesia, with studies from the regional capitals of Medan (2 studies) and Semarang (1 studies), and Malaysia, with studies from the national capital of Kuala Lumpur (2 studies) and the Sabahan state capital Kota Kinabalu (1 study), dominate Asian research into the links between urban livability and PGI (Figure 3) and lift Asia ahead of North America for this type of research (Figure 4). As previously mentioned, detailed information about the geographical distribution of research into the links between quality PGI and urban livability can be found in the dataset shared via the Data Descriptor of Simpson and Parker [29].

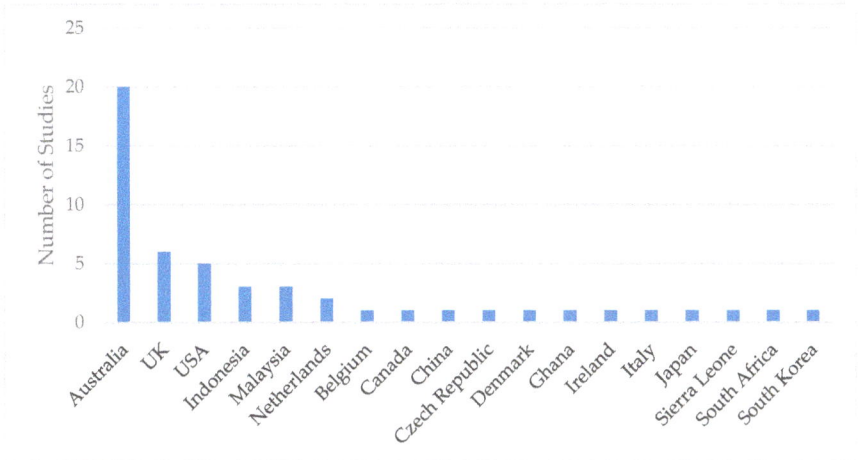

Figure 3. Number of studies that report on linkages between urban livability and PGI spaces for all countries specifically mentioned in the research identified by the initial search. Transnational and international studies reporting on research in multiple countries are recorded against each country.

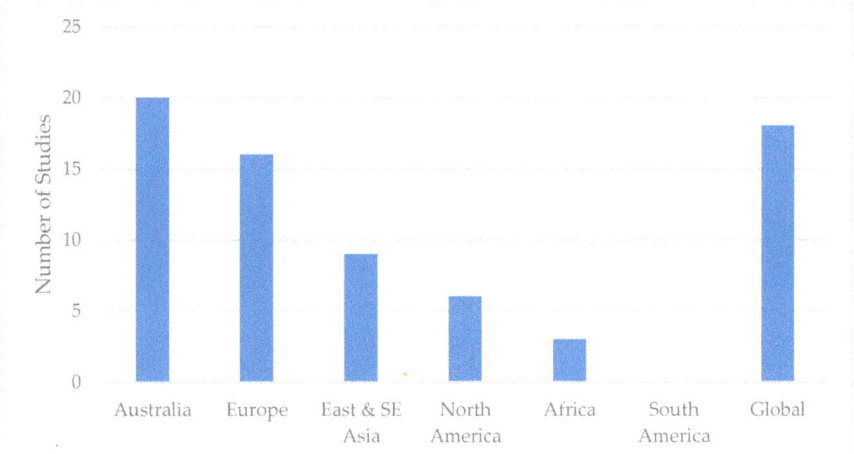

Figure 4. Number of studies that report on linkages between urban livability and PGI spaces by the continent or geographic region identified from the article and the number of globally focused studies (i.e., editorial or review style articles) identified by the initial search. Transnational, international, and continent scale studies are recorded against the relevant continents or geographic regions.

Figure 5. Distribution of research effort into the links between urban livability and PGI identified by the initial search. Markers are placed for all cites specifically mentioned in articles. A marker was not placed for studies where research location was not reported or for articles with a global focus (i.e., editorial or review style articles).

3.3. Reported Research Methods

There is a significant difference ($\chi^2 = 32.57$; $p < 0.001$; $df = 3$) in the rate of reporting of research approaches (Figure 6). A significant majority (Figure 6) of the articles (92%) reported the use of qualitative data collection methods (surveying with open-ended questions, in-depth interviews with participants, observations, and/or focus groups). Slightly fewer articles (73%) reported the use of quantitative data collection methods (observations/recording of frequencies, surveying with scales or closed questions, and computer-generated data), but the rate for that category was also statistically different. Less than half the articles (42%) proposed a new tool or method (new PGI quality assessment tools, new data collection methods, or suggested improvements to existing livability and PGI assessment tools) and a similar number of articles (38%) utilized GIS technology.

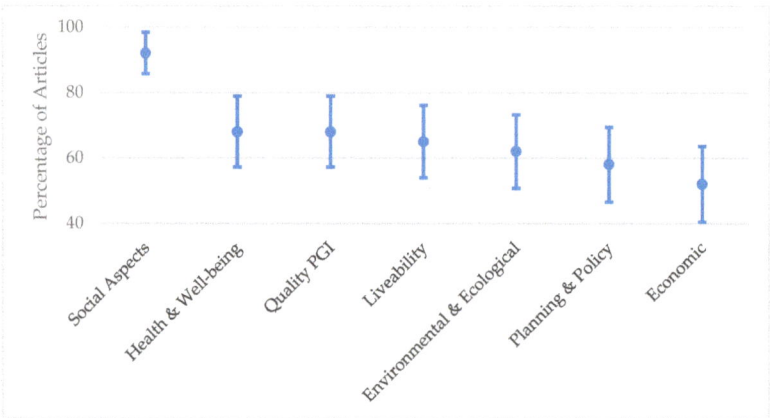

Figure 6. Rate of reporting (± 95% confidence intervals (CIs) of the proportions) of the four research approaches used to investigate the contribution that PGI makes to urban livability identified by this review (n = 71). Percentages add to greater than 100% because the articles that reported a mixed methods approach to research are reported in multiple categories.

The research approaches utilized by each of the articles identified in the initial search of this review are shown in Table 2 and a detailed analysis of the research methods utilized under those four approaches is reported in Table 3. Most articles (86%) reported a mixed methods approach to the research (Table 2).

Table 2. Research approaches utilized, validated, or recommended under each category by study.

Study	GIS (n = 27)	Qualitative (n = 65)	Quantitative (n = 52)	PNT (n = 30)
Antognelli & Vizzari, [48]	✓	✓	✓	✓
Appiah-Opoku [49]		✓	✓	✓
Balding & Williams [50]		✓	✓	✓
Balram & Dragićevic [18]	✓	✓	✓	✓
Barth et al. [51]			✓	
Battisti [52]		✓		✓
Bennett [53]		✓		
Bratman et al. [19]		✓		✓
Cattell et al. [20]		✓		
Čavić & Beirão [54]		✓	✓	✓
Chen et al. [55]		✓	✓	
Chiesura [56]		✓	✓	
Conteh & Oktay [15]		✓	✓	
Crawford et al. [57]	✓	✓	✓	
Dale & Connelly [58]	✓	✓	✓	
Dallimer et al. [59]		✓		
de Lange et al. [60]		✓	✓	
De Riddera et al. [61]	✓		✓	✓
Dietsch et al. [62]			✓	
Do et al. [63]		✓	✓	
Edwards et al. [64]	✓		✓	✓
Francis et al. [65]	✓	✓	✓	
Francis et al. [66]	✓	✓	✓	✓
Gelissen [67]		✓	✓	
Giap et al. [7]	✓	✓	✓	✓
Giles-Corti et al. [68]	✓	✓	✓	✓
Grose [21]	✓	✓	✓	✓
Hagerman [69]		✓		
Hartig et al. [70]		✓		
Hausmann et al. [71]		✓		✓
Hillsdon et al. [72]	✓	✓	✓	
Hock Teck et al. [73]	✓		✓	
Horan et al. [74]	✓	✓	✓	
Howley et al. [75]		✓	✓	
Hughes [22]		✓		
Ikin et al. [76]		✓	✓	
Irvine et al. [77]		✓	✓	
Jones & Newsome [26]	✓	✓	✓	
Kaźmierczak [78]		✓	✓	
Keniger et al. [16]		✓	✓	
Kurniawati [79]		✓		✓
Malek et al. [80]		✓		✓
Manfredo et al. [81]		✓	✓	
Massey [82]		✓	✓	
Nasution & Zahrah [83]		✓		
Nasution & Zahrah [23]		✓		
Newton [27]		✓	✓	✓
Okulicz-Kozaryn [28]		✓		✓
Revell & Anda [84]		✓		

Table 2. *Cont.*

Study	GIS (n = 27)	Qualitative (n = 65)	Quantitative (n = 52)	PNT (n = 30)
Schipperijn et al. [85]	✓	✓	✓	
Schneider & Lorencová [86]		✓	✓	
Shackleton et al. [87]		✓	✓	
Shamsuddin et al. [88]		✓		✓
Shanahan et al. [89]	✓	✓	✓	✓
Shanahan et al. [90]	✓	✓	✓	
Simpson & Newsome [17]		✓		
Soga et al. [91]	✓	✓	✓	
Staats et al. [92]		✓	✓	
Stanley et al. [93]		✓	✓	✓
Sugiyama et al. [94]	✓	✓		✓
Sushinsky et al. [95]	✓	✓	✓	✓
Taylor et al. [96]	✓	✓	✓	✓
Thompson [12]	✓	✓	✓	✓
Tonge & Moore [97]		✓	✓	✓
Turner et al. [98]	✓	✓	✓	
Tzoulas et al. [33]		✓	✓	✓
van den Berg et al. [25]	✓	✓	✓	
Van Herzele & Wiedemann [99]	✓	✓	✓	✓
Villanueva et al. [100]	✓	✓	✓	✓
Wetzstein [101]		✓	✓	
Zhang [102]				✓

GIS = Geographic Information System, PNT = Proposed New Tool. Source Simpson and Parker [29].

Table 3. Research approaches utilized, validated, or recommended under each research category by study (Figure 6).

GIS	Qualitative	Quantitative	Proposed New Tool
Various spatial analyses (28 studies)	Audit (1 study) Case Study (1 study) Experiment (1 study) Focus Group(s) (5 studies) Interviews (8 studies) Import.-Performance Analysis (1 study) Modelling (1 study) Observation (3 studies) Physical Response (1 study) Review Article (26 studies) Survey (20 studies)	Audit (1 study) Experiment (1 study) Focus Group(s) (2 studies) Interviews (1 study) Import.-Performance Analysis (1 study) Modelling (2 studies) Observation (17 studies) Physical Response (1 study) Review Article (15 studies) Survey (13 studies)	New tool for PGI measurement or monitoring (13 studies) Improvement in understanding of PGI attributes or PGI user behaviors (13 studies) Improvement to existing livability measurement or monitoring tool (3 studies) Improvement to existing PGI measurement or monitoring tool (3 studies)

GIS = Geographic Information System.

3.4. Focuses of Urban PGI and Livability Research

It appears that there was a significantly greater research focus with respect to how the social aspects of PGI spaces, such as a sense of community, social needs, social issues, social services, and the human dimensions contribute to urban livability (Figure 7). There was, however, no overall difference in the rate of reporting for the seven research focus areas relating to the contribution that PGI spaces make to urban livability ($\chi^2 = 10.04$; $p = 0.1229$; df = 6). In addition to the 92% of articles identified in the initial search that reported on the contribution of the social aspects of the PGI spaces to urban livability, approximately two thirds of the reviewed articles reported research focuses relating to: human health and well-being aspects (68%); quality of PGI spaces (68%); the contribution that those spaces make to urban livability (65%); and the environmental and/or ecological values of those spaces. (62%). Slightly fewer articles reported on the planning and/or policy aspects of PGI (58%) and economic benefits of PGI in the urban environment (52%).

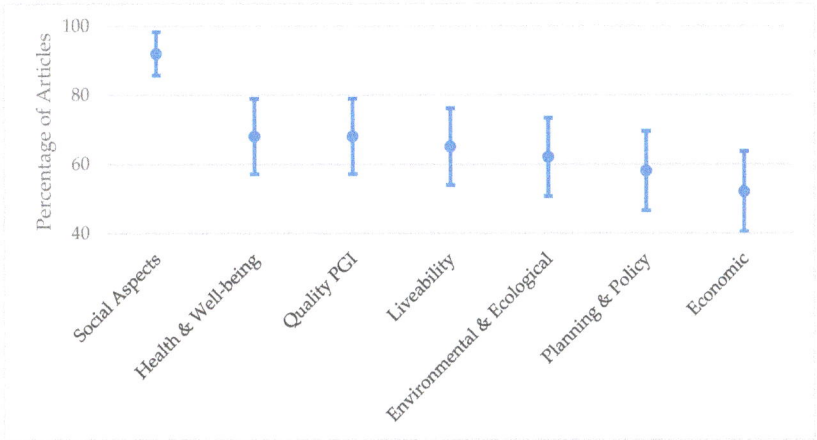

Figure 7. Rate of reporting (± 95% confidence intervals (CIs) of the proportions) of the research focus for the contribution that PGI makes to urban livability identified by this review (n = 71). Percentages add to greater than 100% as articles that report multiple focuses are reported in more than one category.

Articles reported between two and seven focus areas for their PGI-livability research (Figure 8). As for the overall research effort, there was no discernible pattern in the number of research focus areas reported in the articles (χ^2 = 6.584; p = 0.2534; df = 5).

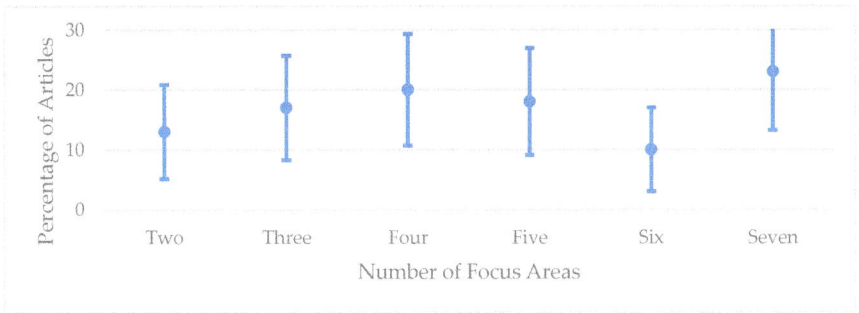

Figure 8. The research focus areas (± 95% CIs of the proportions) reported by articles identified in the initial search reported by this review (n = 71).

Individually collating the research focuses reported in the reviewed articles (Table 4) against the three primary indicators for the contribution of PGI spaces to urban livability (i.e., human health and well-being, livability in its own right, and the social aspects of PGI) also provided no evidence for significant differences in research focus (Figure 9).

Table 4. Percentage of articles reporting on the research focus for articles identified in the initial search of this reviewer that reported on the contribution that PGI makes to urban livability (n = 71). Percentages add to greater than 100% as articles that report multiple research focus areas are reported in more than one category.

Primary Indicators of Livability	Research Focus Areas						
	Health/Wellbeing	Livability	Social Aspects	Quality PGI	Environ./Ecological	Planning/Policy	Economic
Health/Wellbeing Focus	NA	77	90	83	58	63	52
Livability Focus	80	NA	93	72	63	74	61
Social Aspects Focus	66	66	NA	69	65	62	58

NA = Not Applicable as that indicator of livability was the research focus used to cluster the articles.

The high degree of overlap of the 95% CIs of the proportions, shown in Figure 9, were confirmed by the chi-squared analyses for the rate of reporting research focuses aligned to human health and well-being (χ^2 = 7.65; p = 0.1766; df = 5), aligned to urban livability (χ^2 = 4.47; p = 0.4838; df = 5), and aligned to the social aspects of PGI spaces (χ^2 = 7.37; p = 0.9809; df = 5).

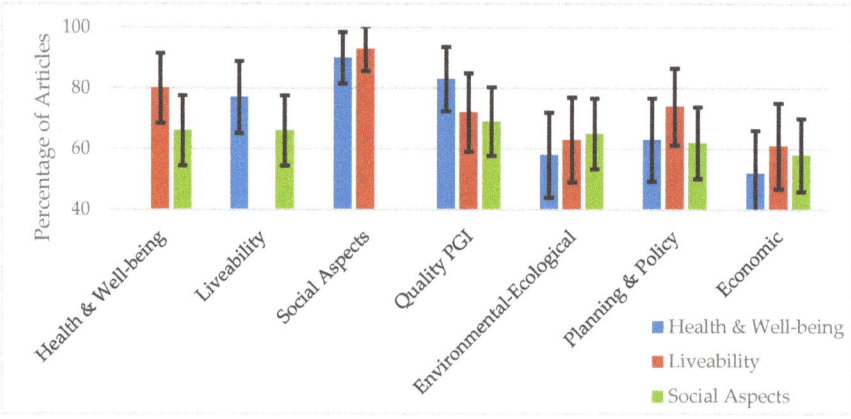

Figure 9. The collated rates of reporting (± 95% CIs of the proportions) of research focuses for the contribution that PGI spaces make to urban livability with respect to the key indicators of human health and well-being, livability in its own right, and the social aspects of PGI spaces. Percentages add to greater than 100% as articles reporting multiple focuses reported in more than one category.

While there is no statistical evidence for any differences in the rate of reporting with respect to livability related factors, nor with respect to the rate of reporting for any particular focus in relation to the three indicators of livability, there is, however, strong evidence of correlations in relation to the reporting of the focuses for research into the contribution that PGI spaces make to urban livability (Table 5). In decreasing order for the strength of the correlation, significant relationships exist with regard to articles identified in the initial search reporting on:

- Human health and wellbeing and the quality of PGI (r = 0.48; p = <0.0001).
- Livability and the planning and policy related to PGI spaces (r = 0.44; p = 0.0001).
- Human health and wellbeing in conjunction with urban livability (r = 0.37; p = 0.0014).
- Social aspects of PGI and economic factors (r = 0.32; p = 0.0071).
- Social aspects of PGI and the planning and policy related to PGI spaces (r = 0.25; p = 0.0335).

It is important to exercise caution when drawing inferences for statistical relationships that have p-values close to the level of significance (α = 0.05), such cases require informed judgment based on the evidence. The correlation between the reporting of urban livability and economic factors (r = 0.24) is comparable to the significant correlation for the social aspects of PGI and the planning and policy related to PGI spaces, so the p = 0.0459 is likely to be supportive of a significant correlation in the reporting of livability and economic factors. In contrast, the r = 0.08 correlation between environmental and ecological values of PGI spaces and the quality of PGI spaces suggests that only about 0.5% of the variability for the reporting of each of those focuses is explained by the reporting of the other. On that basis, the p-value associated with that relationship of 0.472, which rounds to the α =0.05 significance level, cannot be considered to provide evidence of a significant correlation.

Table 5. Correlations (lower left) for research focuses reported in articles and p-values for significance of correlations (upper right). Significant correlations and the associated p-value are indicated by an *.

	Health/Wellbeing	Livability	Social Aspects	Quality PGI	Environ./Ecological	Planning/Policy	Economic
Health/Wellbeing		0.0014	0.3968	< 0.0001*	0.3688	0.2476	0.9944
Livability	0.37*		0.4352	0.3196	0.8042	0.0001*	0.0459
Social Aspects	−0.10	0.09		0.3426	0.1348	0.0335*	0.0071*
Quality PGI	0.48*	0.12	0.11		0.5195	0.2476	0.6128
Environ./Ecological	−0.10	0.03	0.18	0.08		0.4381	0.0472
Planning/Policy	0.14	0.44*	0.25*	0.14	0.09		0.0062*
Economic	<−0.01	0.24	0.32*	−0.06	0.24	0.32*	

3.5. Contributors to Urban Livability

Informed by the research reported in the Data Descriptor of Simpson and Parker [29], research into the linkages between PGI and urban livability identified the six attributes of PGI, shown in Figure 10 and listed in Table 6 as contributing to improved livability of urbanized landscapes. At least one of these factors, and generally more, were reported in 50 of the 71 of the articles (70%) identified in the initial search. The following analyses relate specially to the 50 articles that reported on the contribution that these PGI attributes make to urban livability.

There was a statistically significant difference (χ^2 = 24.79; p < 0.001; df = 5) with respect to the rate of reporting for these PGI attributes in articles identified in the initial search. The quality of PGI spaces was the most reported attribute (84%) with respect to its contribution to urban livability. That rate of reporting of that attribute was significantly greater than for the other five attributes. The rate of reporting of the contribution to urban livability arising from the opportunity to experience environmental and ecological processes in the urban landscape (60%), the presence of PGI spaces in the urban fabric (48%), and the ease of access to those PGI spaces (48%) were all reported at statistically similar rates, as demonstrated by the overlapping 95% CIs, shown in Figure 10. Only mentioned in approximately a quarter of the articles that investigated these six PGI attributes, the walkability of PGI spaces (28%), and the presence of tree canopy cover (24%) were reported at significantly lower rates than the other four attributes.

As for the previously reported focuses of research into the contribution of PGI spaces to urban livability, there were also correlations between the reporting of the six attributes of PGI spaces that this review identifies as contributing to urban livability (Table 6). In decreasing order for the strength of the correlation, the following significant relationships exist with regard to articles identified in the initial search that report on the six attributes of PGI that contribute to urban livability:

- Ease of access to the PGI space and walkability of the PGI space (r = 0.56; p = < 0.0001).
- Walkability of the PGI space and the presence of a tree canopy (r = 0.48; p = 0.0004).
- Presence of a tree canopy and presence of PGI spaces (r = 0.40; p = 0.0043).
- Presence of a tree canopy and ease of access to the PGI space (r = 0.40; p = 0.0043).
- Presence of PGI spaces and walkability of the PGI space (r = 0.38; p = 0.0063).
- Quality of PGI spaces and ease of access to the PGI space (r = 0.31; p = 0.0284).

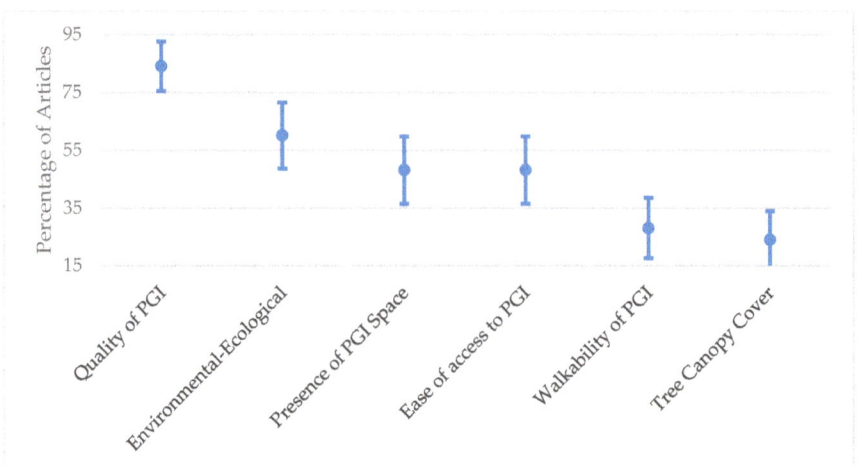

Figure 10. Rate of reporting (± 95% CIs of the proportions) regarding the attributes of PGI identified by this review as contributing to urban livability (n = 50). Percentages add to greater than 100% as articles that report multiple attributes are reported in more than one category.

Table 6. Correlations (lower left) between reporting of PGI attributes that contribute to urban livability and the p-values for significance of the correlations (upper right). Significant correlations and the associated p-value are indicated by an *.

	Quality of PGI	Enviro. & Ecological	Presence of PGI	Ease of Access	Walkability of PGI	Tree Canopy
Quality of PGI		0.3549	0.1618	0.0284*	0.2963	0.4163
Enviro. & Ecological	−0.13		0.1385	0.8217	0.7067	0.0602
Presence of PGI	0.20	0.21		0.0499	0.0062*	0.0043*
Ease of Access	0.31*	−0.03	0.28		<0.0001*	0.0043*
Walkability of PGI	0.15	0.05	0.38*	0.56*		0.0004*
Tree Canopy	0.12	0.27	0.40*	0.40*	0.48*	

Environ. = Environmental.

The coefficient for the correlation between the reporting of the presence of PGI spaces in urbanized landscapes and ease of access to those PGI spaces is only marginally lower than the correlation between the reporting of the quality of PGI spaces and ease of access to those PGI spaces. The p-value for the significance of that correlation (p = 0.0499), however, rounds to the level of significance (α = 0.05), meaning it would be statistically hazardous to draw conclusions about a relationship in the reporting of those two attributes of PGI spaces.

The reporting of these six attributes of PGI spaces that were identified as contributing to urban livability by articles identified in the initial search of this review are shown in Table 7.

Table 7. Reporting of research into six attributes of PGI spaces that contribute to urban livability.

Study	Quality of PGI n = 42	Enviro. & Ecological n = 30	Presence of PGI n = 24	Ease of Access n = 24	Walkability of PGI n = 14	Tree Canopy n = 12
Antognelli & Vizzari [48]	✓	✓	✓	✓		
Balding & Williams [50]		✓				
Barth et al. [51]		✓				
Battisti [52]	✓	✓				
Cattell et al. [20]	✓	✓	✓			
Čavić & Beirão [54]	✓					
Chen et al. [55]	✓					
Chiesura [56]	✓	✓				
Dale & Connelly [58]	✓	✓	✓			
Dallimer et al. [59]		✓				
De Riddera et al. [61]		✓		✓	✓	
Dietsch et al. [62]		✓				
Do et al. [63]	✓	✓				
Francis et al. [65]	✓					
Francis et al. [66]	✓		✓	✓		
Giap et al. [7]	✓	✓	✓	✓	✓	✓
Giles-Corti et al. [68]	✓	✓	✓	✓	✓	✓
Grose [21]	✓	✓	✓	✓	✓	✓
Hagerman [69]	✓	✓				
Hartig et al. [70]	✓			✓		
Hausmann et al. [71]	✓	✓				
Howley et al. [75]	✓		✓			
Hughes [22]	✓	✓	✓	✓		✓
Irvine et al. [77]	✓	✓	✓			
Jones & Newsome [26]	✓	✓	✓	✓	✓	
Kaźmierczak [78]	✓	✓		✓	✓	
Keniger et al. [16]	✓					
Kurniawati [79]	✓			✓		
Nasution & Zahrah [83]	✓		✓			
Nasution & Zahrah [23]	✓					
Newton [27]			✓			
Okulicz-Kozaryn [28]	✓	✓	✓	✓	✓	✓
Revell & Anda [84]		✓	✓			
Schipperijn et al. [85]	✓			✓		
Shackleton et al. [87]						✓
Shamsuddin et al. [88]	✓				✓	
Shanahan et al. [89]	✓	✓				
Shanahan et al. [90]	✓	✓	✓	✓	✓	✓
Simpson & Newsome [17]	✓	✓	✓			✓
Soga et al. [91]	✓	✓	✓	✓		
Sugiyama et al. [94]	✓			✓	✓	
Sushinsky et al. [95]	✓	✓		✓		✓
Taylor et al. [96]	✓		✓	✓	✓	✓
Thompson [12]	✓			✓		
Turner et al. [98]	✓	✓		✓		
Tzoulas et al. [33]	✓	✓	✓			
van den Berg et al. [25]	✓	✓	✓	✓	✓	✓
Van Herzele & Wiedemann [99]	✓			✓		
Villanueva et al. [100]	✓		✓	✓	✓	
Zhang [102]	✓	✓	✓	✓	✓	✓

Environ. = Environmental. Sourced from Simpson and Parker [29].

3.6. Future PGI and Urban Livability Research

Informed by the systematic review reported in the Data Descriptor of Simpson and Parker [29], the initial search revealed that approximately one third (38%) of articles reported a lack of research with respect to the contribution that PGI spaces make to urban livability. A similar number of articles (37%) made recommendations as to the direction that future research should take. Such information is important when determining the current research and knowledge status of the discipline and future research that is required to progress understanding of benefits that urban dwellers gain from experiencing PGI spaces, and the linkages between PGI spaces and urban livability. The articles on the contribution of PGI to urban livability were published in 44 journals covering a variety of disciplines (Table 8). This demonstrates the multidisciplinary nature of this emerging field of research and policy, as well as the need to derive learnings from existing research that is grounded across a range of disciplines. Knowledge gaps and research opportunities identified from the literature will be explored further in the following theoretical Synthesis.

Table 8. Name of journals and frequency of reporting for research into the contribution that PGI spaces make to urban livability (n = 71).

Name of Journal	Number of Articles	Name of Journal	Number of Articles
Landscape and Urban Planning	12	Academic Position Paper	1
Procedia—Social and Behavioral Sciences	5	Acta Universitatis Agriculturae et Silviculturae Mendelianae Brunensis	1
Conservation Biology	4	Ecological Indicators	1
Journal of Environmental Psychology	3	Ecological Management and Restoration	1
American Journal of Preventive Medicine	2	Edited Book	1
American Journal of Public Health	2	Environment and Behavior	1
Applied Geography	2	Environmental Conservation	1
BioScience	2	Frontiers in Ecology and the Environment	1
Wetlands Ecology and Management	2	Geo: Geography and Environment	1
World Review of Science, Technology and Sustainable Development	2	Global Change Biology	1
International Journal of Environmental Research and Public Health	1	Public Health	1
International Journal of the Commons	1	Science of the Total Environment	1
International Journal of Tourism Cities	1	Social Indicators Research	1
Journal of Environmental Planning and Management	1	Social Science & Medicine	1
Journal of Urban Technology	1	Society and Natural Resources	1
Landscape Ecology	1	Sustainability	1
Local Environment	1	Tourism Management	1
New York Academy of Sciences	1	Town Planning Review	1
Open House International	1		

3.7. Outcomes of Supplementary Search

The 16 papers collected during the supplementary search, only 18-months after the initial search, revealed a possible change in the focus of PGI-livability research in the period 2014 to December 2017 inclusive (Figure 11). Only one of the articles from the supplementary search reported specifically on livability as a research focus and only two articles focused on the quality of PGI spaces, hence those two research focuses could not be reliably compared to outcomes of the initial search and have been excluded from the following analysis. While the smaller sample size of articles identified in the supplementary search (n = 16) means that care is needed in interpreting possible changes in the focuses for research, using proportions (percentages) for the chi-square analyses and the wider 95% CIs (Figure 11) compensate for the smaller sample size.

A 2x2 chi-square analysis provided evidence (χ^2 = 73.50; p <<< 0.001; df = 1) that the apparent reduction in the rate at which articles from the supplementary search (31%) recommended a new tool could be a significant trend (42% in the initial search).

Similarly, there is statistical evidence (χ^2 = 15.72; p <<< 0.034; df = 4) that the differences in the rates of reporting of research focuses, shown in Figure 11, may also contain significant trends. With the exception of economic factors, the rate of reporting for all other research focuses appears to have declined between the initial and supplementary searches. With no overlap in the 95% CIs, there is strong evidence for the half as many articles reported research focused on the human health and well-being benefits provided by PGI spaces in the supplementary search (31%) compared to the initial search (68%) being an evolving trend. The research into the social benefits of PGI demonstrated the smallest decline in reporting (18%) between the supplementary (75%) and initial (92%) searches. Research into the contribution that the environment and ecological values of PGI spaces make to urban livability appears to have fallen by approximately one third between the supplementary (44%) and the initial search (62%). There appears to have been a similar one third reduction for research into the links between planning and policy of PGI spaces and livability between the supplementary

(38%) and initial (58%) searches. In contrast to the trends for the other research focuses, shown in Figure 11, the reporting of research into the economic factors and the contribution of PGI spaces to urban livability apparently increased by one third between the initial (52%) and supplementary search (69%). These findings are explored further in the Synthesis of this review.

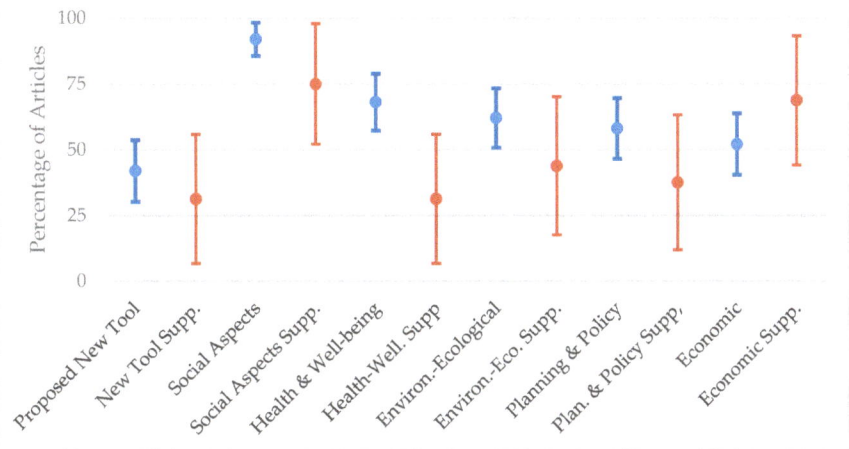

Figure 11. Comparison of changes in reporting rates (± 95% CIs of proportions) for research focuses identified in initial (Blue Data) and supplementary (Dark Red) searches. Supp. = Supplementary, Well. = Well-being, Environ. = Environmental, Eco. = Ecological, and Plan. = Planning.

4. Synthesis

4.1. Quantitative Review Summary

Many PGI researchers report that understanding user experiences, expectations, and satisfaction levels whilst visiting a PGI space is of great value to land managers (e.g., [34,103–107]). Meeting user needs and desires (physical, psychological, and spiritual), as well as providing abundant social, economic, and environmental opportunities, is a primary purpose of urban PGI [103]. Creating and enhancing the synergy between PGI users and land managers is necessary to improve the focus and implementation of management actions and justify the allocation of resources. The primary goal of this review was to provide a plain language overview of recent literature that reports on the psychological, physiological, general well-being, and wider societal benefits that humans receive as a result of experiencing quality PGI spaces in urbanized landscapes and how those PGI spaces contribute to urban livability to improve the quality of life experienced by urban dwellers. The key findings of the quantitative systemic review can be summarized as follows.

Research interest in the benefits that accrue from urban PGI spaces and their contribution to urban livability is growing rapidly with the past six years spawning a 122% increase in published research compared to the first twelve years of this century (Figure 2). While the majority of that research was centered on mid-latitude cities located in the developed nations of Australia, Europe, and North America (Figure 4), research from Indonesia and Malaysia is also making a significant contribution to understanding how PGI spaces contribute to urban livability. The two global hotspots for this research are the southeast of the United Kingdom and Perth, the geographically isolated state capital located in the southwest of Western Australia (Figure 5).

Approximately four out of ten studies identified in the initial search conducted under this systematic review reported on tools that could be used to assess or quantify the quality of PGI spaces and the contribution those spaces made to urban livability (Figure 6). A similar number of articles analyzed spatial data to explore the characteristics of urban PGI spaces (Figure 6). The majority of the research, however, relied on traditional mixed methods research (Tables 2 and 3) to gather qualitative and quantitative data about PGI spaces and the people who use and manages those spaces. That research was focused on the seven focus areas (Figure 7) of the social aspects of PGI spaces, the human health and well-being benefits that accrue from visiting PGI spaces, the quality of urban PGI spaces, research focused explicitly on urban livability, the environmental and ecological values of PGI spaces, policy and planning issues associated with urban PGI, and economic factors related to PGI spaces.

The quantitate review reported in the Results section of this article found no statistical evidence of a difference with respect to the rate of reporting of these seven research focus areas, but there were statistically significant correlations between the reporting of some of the research focuses (Table 4). There was a significant correlation for articles that linked the reporting of human health and well-being benefits accruing from visiting PGI spaces and research into the quality of PGI spaces. There was also a correlation between articles reporting on urban livability in combination with research of the planning and policy related to PGI spaces. Given that almost eight out of tenarticles (Figure 9) that mentioned livability also reported on policy and planning, the management framework provided for PGI spaces is an important consideration for urban planners and land managers wishing to optimize the livability of their urban community. The definition of urban livability postulated by Giap et al. [7] and adopted by this review is supported by the positive correlation between the 80% of articles identified in the initial search that reported on both the human health and well-being benefits that accrue from visiting quality PGI spaces and urban livability. There were also significant correlations between articles that reported on both economic factors and planning and policy related to PGI spaces with articles that reported on the social aspects of the PGI spaces. The reporting for those areas of research were, however, lower than other aspects of PGI-livability research, which means the interpretation of those correlations would require a deeper analysis that was beyond the scope of this review.

This review, informed by the data reported in Simpson and Parker [29], identified six attributes of PGI spaces that are reported by the PGI-livability literature as contributing to improved livability of urbanized landscapes. In order of decreasing frequency of reporting (Figure 10), those attributes are the quality of PGI spaces, the opportunities that PGI space provide to experience the natural environmental and ecological processes (i.e., UN), the presence of PGI spaces in the urban fabric, ease of access to PGI spaces (in terms of both availability and location of PGI spaces and PGI spaces being equitably—socially and physically—accessible all community members), the internal walkability of PGI spaces, and the presence of tree canopy cover at PGI sites.

There was a statistically significant difference in the rate of reporting of these attributes by articles identified in the initial search. Reporting on the contribution of quality of PGI spaces to urban livability was significantly greater than for the other five attributes with slightly better than eight out of every ten articles mentioning that PGI attribute. Only mentioned by approximately one in four articles, the contributions that internal walkability of PGI spaces and the presence of a tree canopy at PGI sites make to urban livability were both reported at a rate significantly below the other four PGI attributes that contribute to urban livability.

There were also significant correlations between the reporting of these attributes, but some those correlations may have a different interpretation of the correlations between the areas of research focus summarized above. The strongest correlation related to reporting of the ease of access to PGI spaces and the walkability of the PGI space. The apparently obvious interpretation of this relationship would be that if PGI spaces are easily accessible, then they would also be highly walkable spaces. Caution needs to be exercised however, as the low reporting of the internal walkability of PGI spaces and the binary nature of the data means it is possible that this correlation is significant because there is an underreporting of PGI walkability (28% of articles) while almost half of all articles (49%) reported on the contribution that ease of access to PGI spaces makes to urban livability. It was beyond the primary focus and scope of this review to determine which of those scenarios apply. With similar rates of reporting, it is likely that the presence of tree canopy cover does influence the internal walkability of

PGI spaces and that the contribution that the combination of those attributes makes to city livability means they are commonly reported together in the PGI-livability literature. Intuitively, the correlation between articles reporting on the presence of a tree cover canopy and the presence of PGI spaces in urban landscapes appears self-evident and a statistical relationship between the reporting of those two attributes would appear to have real-world meaning. Once again however, the disparity in the rate of reporting between the presence of a tree canopy (24%) and the presence of PGI spaces (48%), means this correlation could also reflect a significant underreporting of the contribution that a tree canopy makes to urban livability. An interpretation of the correlation between the reporting on the presence of a tree cover canopy at PGI spaces and the ease of access to PGI spaces is not immediately apparent, and similar to the correlation between reporting of the ease of access to PGI spaces and the walkability of the PGI space, further investigation of the literature needs to understand the relationship between the reporting of these attributes. As for the other correlations to reporting of the PGI walkability attribute, the correlation with reporting on the presence of PGI spaces (48%) could also reflect the under reporting of walkability (28%) in the PGI-livability literature. Similarly, the contribution that quality of PGI spaces make to urban livability is reported in significantly more articles than the ease of access to PGI spaces (84% compared to 48%), so further investigation, which was beyond the scope of this review, would be needed to interpret the real-world significance of the correlation between the reporting of those two PGI attributes.

It was noted during the quantitative analysis that articles from the supplementary search appeared more focused in their research and reported on fewer research topics than the articles obtained in the initial search, which may have contributed to lower levels of reporting of both new tools for assessing the contribution of PGI spaces to urban livability and the benefits to human health and well-being that accrue from visiting PGI spaces. To increase confidence that these apparent differences between the initial and supplementary searchers are true trends, further monitoring of the literature will be required to complement the small number of new articles identified in the supplementary search.

4.2. PGI and Livability

Living in highly urbanized environments often results in diminished opportunities to experience nature. Researchers, such as Soga et al. [91], advocate for the need to increase the value placed on quality PGI spaces, particularly in highly urbanized areas, to reduce the disengagement of people with the surrounding natural world.

A system of quality PGI that supports indigenous ecosystems and sustainable ecological processes is a key determinant for the livability of urbanized landscapes [7]. As demonstrated in the Results section and summarized above, investigating the contribution that the environmental and ecological values of UN make to urban livability has been a key focus for researchers. With environmental and ecological values demonstrated to be a key contributor to improving the livability of urbanised landscapes, urban planners and managers should work to protect conserve, and renaturing PGI spaces for the betterment of their communities and cities more broadly.

When exposed to UN in a quality PGI space, humans experience a greater sense of well-being with psychological, physiological, and biological factors all contributing [16,18–25]. As reported by Giap et al. [7] and reinforced by this review, human well-being is a primary indicator for levels of urban livability. During this time when humans are highly connected via rapidly progressing technology and experiencing highly demanding and competitive working environments, capitalizing on opportunities to engage with the natural environment for psychological benefit has enormous potential for improving social health and well-being outcomes [108], thus making urbanized landscapes more livable. Engaging with elements of the natural environment allows individuals to connect deeply within ourselves, with others, to experience wonder, and to be organically inspired; all representing a marked psychological change for the better [108].

As was highlighted in the Introduction, research into physiological benefits for individuals who engage with the natural environment is also an area of rapid progression. An increase in physical activity, improved physical fitness, improved cardiovascular health, and an improvement in children's agility and spatial negotiation skills are just some of the documented benefits [16,23,109]. Researchers, such as Gladwell et al. [110] and Li et al. [111], advocate that exposure to nature and nature experiences sees a reduction in stress, improvements in mood, assists in restoring mental fatigue, and enhances the perception of our own physiological health. Those findings are likely to underpin the strong research focus on human health and well-being, quality of PGI spaces, and contribution to urban livability that arise from humans experiencing environmental and ecological processes in PGI spaces reported in this review. The reports of Gladwell et al. [110] and Li et al. [111] and the findings of our review explain why the presence of quality PGI spaces that contain, and support, UN are essential for optimizing the livability of urbanized landscapes.

In established cities with limited opportunities for the acquisition of new land for PGI, creative thinking needs to be applied if adequate PGI spaces are to be provided to residents. Informal PGI installations, such as pop-up gardens, parklets, roof top gardens, and green walls can provide the physical and psychological well-being markers similarly afforded by the traditional PGI spaces that were the focus of this review [2,112]. Examples of successful alternative PGI installations, in line with the Biophilic Hypothesis, made famous by Wilson [1], include an acclaimed green wall at the Musée du Quai Branly (close to the Eiffel Tower) installed by the well-known botanist Patrick Blanc [2] covering the entire façade of a three-story building. Passersby are observed to stop to admire, gaze, and stand in awe of this spectacle [2]. Another is the rooftop garden of the Ballard branch of the Seattle Public Library, which comprises around eighteen thousand native shrub and grass species, which has proven successful in passive heating and cooling, contributing largely to local biodiversity, and acting as an educational showcase for residents and visitors to the area [2]. With rising populations and intensifying density, creative thinking can assist in achieving a way forward to harness diverse opportunities to provide UGI. In addition, PGI resourcing and design is intrinsically linked to the demographics of a population and city. For example, current trends in Australia are showing that people are choosing to have fewer children, have children later in life, and are living longer, due to medical and technological advancements [112]. The combination of these factors is resulting in a significantly aging population [112]. Pre-empting the needs of an aging population means assessing the accessibility and safety of PGI spaces, providing the required infrastructure to support older visitors, creating and supporting passive spaces, and planning for different levels of mobility [112].

While resource allocation is a circular debate, ongoing and adequate resources are required to meet the needs of human populations living in urbanized landscapes. When determining the *quality* in quality PGI, the following seven areas are consistently reported in current PGI literature: functionality, fair and equitable access, conservation and environmental education, water sensitive management, meeting social needs, infrastructure, and amenities [48,68,77,89,100,113]. The level of performance across these features gives an indication of the overall quality of a PGI space. Examples of high performance among these features include: consistent universal access across the site—including infrastructure (e.g., pathways, picnic tables, and playgrounds); use of water sensitive turf and plants; and a practicable site layout with installations that meet the current needs of PGI users. Each of the seven areas above contributes to the visitor experience of a PGI space. It is generally agreed that these features create a foundation of the visitation experience of a typical urban PGI space [48,68,77,89,100,113].

4.3. Livability Ranking Scales

Urban livability rankings and performance may be considered by different countries for a variety of reasons. An example is Auckland, New Zealand, where an Urban Growth Management Strategy strongly linked to, and influenced by, urban livability and quality of life considerations are currently being pursued [103]. Current livability scales can reveal information about the performance (high and low) of critical elements of urban centers, which can inform many aspects of urban design and management, including ongoing debates around resource allocation. Research has now progressed into understanding the trade-offs that residents may be willing to make when choosing where to live [103]. It has been found that there may be a willingness to forgo quality public transport options to live farther from the city center on a larger parcel of private land, or similarly a willingness to pay a premium price to live in a more central location. This is known as the tradeoff between suburban and urban lifestyles [103]. Understanding how these tradeoffs affect the perception of quality of life and livability is necessary to assess, interpret, and enhance the efficacy and quality of such livability scales [103]. Research into the amenities available to each lifestyle is a largely under researched area. A better understanding in this area would bring value to decision making processes, provision of services, and qualifying resident values in developed, developing, and less developed nations [103]. Ultimately, a progression in such knowledge would assist in the balanced approach to providing PGI that meets the lifestyle choices and expectations and urban dwellers, and would deliver a range of PGI spaces that are perceived to be high value to residents within a city, helping to achieve a diverse, highly functioning, and quality urban living experience. The quantitative analyses and theoretical synthesis reported in this review and the Importance-Performance Analysis case study presented in Parker and Simpson [35] provide the background information and demonstrate an assessment method that will allow urban planners, land managers, and other stakeholders to assess the quality of PGI spaces in their locale and to make informed decisions that will enhance the contribution that those spaces make to both the livability of their community and the broader urban landscape of their city. Citizens who perceive a city to be highly livable are more likely to engage, experience, and enjoy the benefits that the city can offer [26,28]. Understanding the factors that contribute to the *perception* of urban livability is essential for a true and equitable concept of the city. As previously reported in this review, the key determinants of the contribution that PGI spaces can make to the livability of urban landscapes reported in the literature are the presence of quality PGI spaces, that are easily accessed, highly walkable internally, that have tree canopy cover, and, perhaps most importantly, provide the opportunity to experience and engage with quality environmental and ecological systems (i.e., UN), which is an innate need of humanity.

Certain limitations exist in the current livability scales [114]. The current 'one size fits all' ethos for city planning, which includes the provision of PGI and retention of UN, does not consider the development stages of a city when it is under-developed, developing, or developed [114]. Currently, subjective elements, such as opportunities for nature experiences, environmental education, and opportunities for visiting quality PGI are under-represented in livability scales [26]. This may be due to the poor understanding of these elements, the difficulty in quantifying and assigning a 'score' for PGI, and/or the difficulty in verifying 'performance' of urban PGI and UN spaces. The subjective elements that may contribute to urban livability could include how a city protects fragile ecosystems, responds to climate change, funds environmental education programs, addresses resource recovery and waste minimization, the degree of resource depletion, and the social value placed on leisure time [114]. It is the synthesis of this paper, based on the systematic review and guidance of current literature, that the presence and prevalence of high quality PGI and UN (remnant and renatured) is a strong contributor to an individuals' perception of urban livability.

4.4. Knowledge Gaps and Future Research

Informed by this review and the Data Descriptor of Simpson and Parker [29], several research gaps pertaining to the different aspects of PGI spaces, the contributions that PGI spaces make to urban livability, and the contribution of PGI to the general health and well-being of urbanized human populations emerged. The research suggested below is relevant for the disciplinary progression, increased research legitimacy, and better provisioning and servicing of urban PGI spaces. Additional research regarding linkages between PGI and urban livability should be focused on:

1. Measures to achieve greater consistency and consensus with respect to terminology, measurement methods, land management approaches, and policy development related to PGI and urban livability.

2. Investigating the correlations in the rate of reporting of PGI attributes that contribute to urban livability to determine if there are real-world explanations for the patterns identified in the literature or if the correlations reported in this review arose from the discrepancy in the rate of research and/or reporting with respect to some PGI attributes.

3. Replication of existing research to enhance research integrity, particularly with respect to broadly focused research that will identify quality markers of PGI assets and enhance the contribution that quality PGI spaces make to urban livability.

4. Research to further elucidate why exercising within the natural environment requires lessened exertion when compared to exercising indoors.

5. Enhanced understanding of how PGI assets can increase the resilience of urban centers in a cost effective and socially-centric way.

6. Research regarding how access to quality PGI assets influences the quality of citizen life with respect to the concept of urban livability in developing and less developed nations.

7. Determining what aspects of a local PGI space are important enough to surrounding residents that they are willing to contribute personal resources, such as time (volunteering) and financial donations to enhance the site.

5. Conclusions

While the PGI and UN research that underpins this review was intentionally weighted towards green public open spaces, such as parks and nature conservation areas, the current literature points to PGI and UN being valuable assets that make important contributions to urban livability, which enhances the quality of life for urbanized human communities. These PGI and UN assets are valuable because they provide numerous social, environmental, economic, and health benefits to urban dwellers. Community members, land managers, urban planners, PGI and livability researcher, and other stakeholders who wish to optimize the livability of urbanized landscapes, and consequently the quality of life of within their community, should give due regard to the complementary aspects of PGI spaces reported in this review, specifically the social aspects and benefits of quality PGI spaces, the human health and well-being benefits arising from visiting quality PGI spaces, the opportunities that PGI spaces provide for urban residents to fulfill their innate need to experience and engage with authentic UN spaces, the planning and policy frameworks associated with the provision and management of quality PGI spaces, and the economic costs and benefits that accrue from the provision of quality PGI spaces. When making decisions with respect to the provision and management of PGI spaces, stakeholders need to be mindful of the six attributes of PGI spaces that this review identifies as making the greatest contribution to urban livability. Those attributes are the presence and persistence in urbanized landscape of PGI spaces that incorporate UN, the quality of those PGI spaces, easy and equitable access to PGI spaces both in a physical and social sense, the importance of PGI spaces in providing urban dwellers with the opportunity to experience and engage with healthy functioning indigenous ecosystems, the internal walkability (and we suggest universal access) of PGI spaces, and the need for tree canopy cover at PGI sites.

Supplementary Materials: The following are available online at http://www.mdpi.com/2073-445X/7/4/161/s1, Table S1: Resources utilized in this systematic quantitative literature review.

Author Contributions: J.P. and G.S. made equal contributions to this paper and as such are co-first authors.

Funding: This research received no external funding.

Acknowledgments: We thank our colleague David Newsome for his guidance on the Master's research by J.P. and comments on the associated thesis. We also give thanks to the Guest Editor of the *Land Urbanism and Green Infrastructure* special issue, and two anonymous reviewers whose comments enhanced our article. We would like to give particular thanks to *Land* Editorial Team for their informative, professional, and timely assistance that facilitated the publication of our article.

Conflicts of Interest: The authors declare no conflict of interest.

References

1. Wilson, E.O. *Biophilia*; Harvard University Press: Cambridge, MA, USA, 1984; ISBN 978-0-6740744-2-2.
2. Beatley, T. *Biophilic Cities: Integrating Nature into Urban Design and Planning*; Island Press: Washington, DC, USA, 2011; ISBN 978-1-5972671-5-1.
3. Miller, J.R. Biodiversity conservation and the extinction of experience. *Trends Ecol. Evol.* **2005**, *20*, 430–434. [CrossRef] [PubMed]
4. Neuman, M. The compact city fallacy. *J. Plan. Educ. Res.* **2005**, *25*, 11–26. [CrossRef]
5. Kopecká, M.; Szatmári, D.; Rosina, K. Analysis of urban green spaces based on Sentinel-2A: Case studies from Slovakia. *Land* **2017**, *6*, 25. [CrossRef]
6. Cameron, R.W.; Blanuša, T.; Taylor, J.E.; Salisbury, A.; Halstead, A.J.; Henricot, B.; Thompson, K. The domestic garden—Its contribution to urban green infrastructure. *Urban For. Urban Green.* **2012**, *11*, 129–137. [CrossRef]
7. Giap, T.K.; Thye, W.W.; Aw, G. A new approach to measuring the liveability of cities: The Global Liveable Cities Index. *World Rev. Sci. Technol. Sustain. Dev.* **2014**, *11*, 176–196. [CrossRef]
8. Kondo, M.; Fluehr, J.; McKeon, T.; Branas, C. Urban Green Space and Its Impact on Human Health. *Int. J. Environ. Res. Public Health* **2018**, *15*, 445. [CrossRef] [PubMed]
9. Roy, S.; Byrne, J.; Pickering, C. A systematic quantitative review of urban tree benefits, costs, and assessment methods across cities in different climatic zones. *Urban For. Urban Green.* **2012**, *11*, 351–363. [CrossRef]
10. Swanwick, C.; Dunnett, N.; Woolley, H. Nature, role and value of green space in towns and cities: An overview. *Built Environ.* **2003**, *29*, 94–106. [CrossRef]
11. Taylor, L.; Hochuli, D.F. Defining greenspace: Multiple uses across multiple disciplines. *Landsc. Urban Plan.* **2017**, *158*, 25–38. [CrossRef]
12. Thompson, C.W. Urban open space in the 21st century. *Landsc. Urban Plan.* **2002**, *60*, 59–72. [CrossRef]
13. Norton, B.A.; Coutts, A.M.; Livesley, S.J.; Harris, R.J.; Hunter, A.M.; Williams, N.S. Planning for cooler cities: A framework to prioritise green infrastructure to mitigate high temperatures in urban landscapes. *Landsc. Urban Plan.* **2015**, *134*, 127–138. [CrossRef]
14. Unterweger, P.A.; Schrode, N.; Betz, O. Urban Nature: Perception and Acceptance of Alternative Green Space Management and the Change of Awareness after Provision of Environmental Information: A Chance for Biodiversity Protection. *Urban Sci.* **2017**, *1*, 24. [CrossRef]
15. Conteh, F.M.; Oktay, D. Measuring liveability by exploring urban qualities of Kissy Street, Sierra Leone. *Open House Int.* **2016**, *41*, 23–30.
16. Keniger, L.E.; Gaston, K.J.; Irvine, K.N.; Fuller, R.A. What are the benefits of interacting with nature? *Int. J. Environ. Res. Public Health* **2013**, *10*, 913–935. [CrossRef] [PubMed]
17. Simpson, G.; Newsome, D. Environmental history of an urban wetland: From degraded colonial resource to nature conservation area. *Geo Geogr. Environ.* **2017**, *4*, E00030. [CrossRef]
18. Balram, S.; Dragicevic, S. Attitudes toward urban green spaces: Integrating questionnaire survey and collaborituve GIS techniques to improve attitude measurements. *Landsc. Urban Plan.* **2005**, *71*, 147–162. [CrossRef]
19. Bratman, G.N.; Hamilton, P.; Daily, G.C. The impacts of nature experience on human cognitive function and mental health. *N. Y. Acad. Sci.* **2012**, *1249*, 118–136. [CrossRef] [PubMed]
20. Cattell, V.; Dines, N.; Gesler, W.; Curtis, S. Mingling, observing, and lingering: Everyday public spaces and their implications for well-being and social relations. *Health Place* **2008**, *14*, 544–561. [CrossRef]
21. Grose, M.J. Changing relationships in public open space and private open space in suburbs in south-western Australia. *Landsc. Urban Plan.* **2009**, *92*, 53–63. [CrossRef]
22. Hughes, M. Researching the links between parklands and health. In *Wellness Tourism: A Destination Perspective*; Voigt, C., Pforr, C., Eds.; Routledge: Abingdon, UK, 2014; pp. 147–160, ISBN 978-1-1-380820-0-7.
23. Nasution, A.D.; Zahrah, W. Public Open Space and Quality of Life in Medan, Indonesia. *Procedia Soc. Behav. Sci.* **2014**, *168*, 219–228. [CrossRef]
24. Patroni, J.; Day, A.; Lee, D.; Chan, J.K.L.; Kerr, D.; Newsome, D.; Simpson, G.D. Looking for evidence that place of residence influenced visitor attitudes to feeding wild dolphins. *Tour. Hosp. Manag.* **2018**, *24*, 87–105. [CrossRef]
25. Van den Berg, A.E.; Hartig, T.; Staats, H. Preference for nature in urbanized societies: Stress, restoration, and the pursuit of sustainability. *J. Soc. Issues* **2007**, *63*, 79–96. [CrossRef]

26. Jones, C.; Newsome, D. Perth (Australia) as one of the world's most liveable cities: A perspective on society, sustainability and environment. *Int. J. Tour. Cities* **2015**, *1*, 18–35. [CrossRef]
27. Newton, P.W. Liveable and sustainable? Socio-technical challenged for the twenty-first century cities. *J. Urban Technol.* **2012**, *19*, 81–102. [CrossRef]
28. Okulicz-Kozaryn, A. City life: Rankings (liveability) versus perceptions (satisfaction). *Soc. Indic. Res.* **2011**, *110*, 433–451. [CrossRef]
29. Simpson, G.; Parker, J. Data on Peer Reviewed Papers about Green Infrastructure, Urban Nature, and City Liveability. *Data* **2018**, *3*, 51. [CrossRef]
30. Conger, B. SPP Research Paper No. 7-4: On Livability, Liveability and the Limited Utility of Quality-of-Life Rankings. Available online: https://papers.ssrn.com/sol3/papers.cfm?abstract_id=2614678# (accessed on 24 August 2018).
31. The Value of Rankings and the Meaning of Livability. Available online: http://www.livablecities.org/blog/value-rankings-and-meaning-livability (accessed on 7 August 2018).
32. Quality of Life Survey: Top 25 Cities. 2018. Available online: https://monocle.com/film/affairs/quality-of-life-survey-top-25-cities-2018/ (accessed on 18 November 2018).
33. Tzoulas, K.; Korpela, K.; Venn, S.; Yli-Pelkonen, V.; Kázmierczak, A.; Niemela, J.; James, P. Promoting ecosystem and human health in urban areas using Green Infrastructure: A literature review. *Landsc. Urban Plan.* **2007**, *81*, 167–178. [CrossRef]
34. Parker, J. A Survey of Park User Perception in the Context of Green Space and City Liveability: Lake Claremont, Western Australia. Master's Thesis, Murdoch University, Perth, Australia, 2017. Available online: http://researchrepository.murdoch.edu.au/id/eprint/40856/ (accessed on 8 October 2018).
35. Parker, J.; Simpson, G. Visitor satisfaction with a public green infrastructure and urban nature space in Perth, western Australia. *Land* **2018**, *7*, 159–176. [CrossRef]
36. Simpson, G.; Parker, J. Data for an Importance-Performance Analysis (IPA) of a public green infrastructure and urban nature space in Perth, western Australia. *Data* **2018**, *7*, 69–78. [CrossRef]
37. Pickering, C.M.; Byrne, J. The benefits of publishing systematic quantitative literature reviews for PhD candidates and other early career researchers. *High. Educ. Res. Dev.* **2013**, *33*, 534–548. [CrossRef]
38. Moher, D.; Liberati, A.; Tetzlaff, J.; Altman, D.G. Preferred reporting items for systematic reviews and meta-analyses: The PRISMA statement. *PLoS Med.* **2009**, *6*. [CrossRef] [PubMed]
39. Berenson, M.L.; Levine, D.M.; Krehbiel, T.C. *Basic Business Statistics: Concepts and Applications: International Edition*, 10th ed.; Pearson Prentice Hall: Upper Saddle River, NJ, USA, 2006; p. 273, ISBN 978-0131536869.
40. Cumming, G.; Fidler, F.; Vaux, D.L. Error bars in experimental biology. *J. Cell Biol.* **2007**, *177*, 7–11. [CrossRef] [PubMed]
41. Berenson, M.L.; Levine, D.M.; Krehbiel, T.C. *Basic Business Statistics: Concepts and Applications: International Edition*, 10th ed.; Pearson Prentice Hall: Upper Saddle River, NJ, USA, 2006; pp. 454–474, ISBN 978-0131536869.
42. Bishara, A.J.; Hittner, J.B. Testing the significance of a correlation with nonnormal data: Comparison of Pearson, Spearman, transformation, and resampling approaches. *Psychol. Methods* **2012**, *17*, 399–417. [CrossRef] [PubMed]
43. Havlicek, L.L.; Peterson, N.L. Robustness of the Pearson Correlation against Violations of Assumptions. *Percept. Motor Skills* **1976**, *43*, 1319–1334. [CrossRef]
44. Rodrıguez, G.; Elo, I. Intra-class correlation in random-effects models for binary data. *Stata J.* **2003**, *3*, 32–46. [CrossRef]
45. Demirtas, H.; Amatya, A.; Doganay, B. BinNor: An R package for concurrent generation of binary and normal data. *Commun. Stat.-Simul. Comput.* **2014**, *43*, 569–579. [CrossRef]
46. Edwards, A.L. *Statistical Methods for the Behavioral Sciences*; Holt, Rinehart and Winston: New York, NY, USA, 1962; pp. 301–304.
47. Berenson, M.L.; Levine, D.M.; Krehbiel, T.C. *Basic Business Statistics: Concepts and Applications: International Edition*, 10th ed.; Pearson Prentice Hall: Upper Saddle River, NJ, USA, 2006; pp. 266, 546–547, ISBN 978-0131536869.
48. Antognelli, S.; Vizzari, M. Landscape liveability spatial assessment integrating ecosystem and urban services with their perceived importance by stakeholders. *Ecol. Indic.* **2017**, *72*, 703–725. [CrossRef]

49. Appiah-Opoku, S. Using protected areas as a tool for biodiversity conservation and ecotourism: A case study of Kakum National Park in Ghana. *Soc. Nat. Resour.* **2011**, *24*, 500–510. [CrossRef]

50. Balding, M.; Williams, K.J. Plant blindness and the implications for plant conservation. *Conserv. Biol.* **2016**, *30*, 1192–1199. [CrossRef]

51. Barth, B.J.; FitzGibbon, S.I.; Wilson, R.S. New urban developments that retain more remnant trees have greater bird diversity. *Landsc. Urban Plan.* **2015**, *136*, 122–129. [CrossRef]

52. Battisti, C. Experiential key species for the nature-disconnected generation. *Anim. Conserv.* **2016**, *19*, 485–487. [CrossRef]

53. Bennett, N.J. Using perceptions as evidence to improve conservation and environmental management. *Conserv. Biol.* **2016**, *30*, 582–592. [CrossRef] [PubMed]

54. Čavić, L.; Beirão, J.N. Open Public Space Attributes and Categories—Complexity and Measurability. *Arhit. Raziskave* **2014**, *2*, 15–24.

55. Chen, B.; Adimo, O.A.; Bao, Z. Assessment of aesthetic quality and multiple functions of urban green space from the users' perspective: The case of Hangzhou Flower Garden, China. *Landsc. Urban Plan.* **2009**, *93*, 76–82. [CrossRef]

56. Chiesura, A. The role of urban parks for the sustainable city. *Landsc. Urban Plan.* **2004**, *68*, 129–138. [CrossRef]

57. Crawford, D.; Timperio, A.; Giles-Corti, B.; Ball, K.; Hume, C.; Roberts, R.; Andrianopoulos, N.; Salmon, J. Do features of public open spaces vary according to neighbourhood socio-economic status? *Health and Place* **2008**, *14*, 889–893. [CrossRef]

58. Dale, P.E.R.; Connelly, R. Wetlands and human health: An overview. *Wetl. Ecol. Manag.* **2012**, *20*, 165–171. [CrossRef]

59. Dallimer, M.; Irvine, K.N.; Skinner, A.M.; Davies, Z.G.; Rouquette, J.R.; Maltby, L.L.; Warren, P.H.; Armsworth, P.R.; Gaston, K.J. Biodiversity and the feel-good factor: Understanding associations between self-reported human well-being and species richness. *BioScience* **2012**, *62*, 47–55. [CrossRef]

60. De Lange, E.; Woodhouse, E.; Milner-Gulland, E.J. Approaches used to evaluate the social impacts of protected areas. *Conserv. Lett.* **2016**, *9*, 327–333. [CrossRef]

61. De Riddera, K.; Adamecb, V.; Bañuelosc, A.; Brused, M.; Bürgerd, M.; Damsgaarde, O.; Dufekb, J.; Hirschf, J.; Lefebrea, F.; Pérez-Lacorzanac, J.M.; et al. An integrated methodology to assess the benefits of urban green space. *Sci. Total Environ.* **2004**, *334–335*, 489–497. [CrossRef]

62. Dietsch, A.M.; Teel, T.L.; Manfredo, M.J. Social values and biodiversity conservation in a dynamic world. *Conserv. Biol.* **2016**, *30*, 1212–1221. [CrossRef]

63. Do, Y.; Kim, S.B.; Kim, J.Y.; Joo, G.J. Wetland-based tourism in South Korea: Who, When, and Why. *Wetl. Ecol. Manag.* **2015**, *23*, 779–787. [CrossRef]

64. Edwards, N.; Hooper, P.; Trapp, G.S.; Bull, F.; Boruff, B.; Giles-Corti, B. Development of a public open space desktop auditing tool (POSDAT): A remote sensing approach. *Appl. Geogr.* **2013**, *38*, 22–30. [CrossRef]

65. Francis, J.; Giles-Corti, B.; Wood, L.; Knuiman, M. Creating sense of community: The role of public space. *J. Environ. Psychol.* **2012**, *32*, 401–409. [CrossRef]

66. Francis, J.; Wood, L.J.; Knuiman, M.; Giles-Corti, B. Quality or quantity? Exploring the relationship between Public Open Space attributes and mental health in Perth, Western Australia. *Soc. Sci. Med.* **2012**, *74*, 1570–1577. [CrossRef] [PubMed]

67. Gelissen, J. Explaining popular support for environmental protection: A multilevel analysis of 50 nations. *Environ. Behav.* **2007**, *39*, 392–415. [CrossRef]

68. Giles-Corti, B.; Broomhall, M.H.; Knuiman, M.; Collins, C.; Douglas, K.; Ng, K.; Lange, A.; Donovan, R.J. Increasing walking: How important is distance to attractiveness, and size of public open space? *Am. J. Prev. Med.* **2005**, *28*, 169–176. [CrossRef]

69. Hagerman, C. Shaping neighborhoods and nature: Urban political ecologies of urban waterfront transformations in Portland, Oregon. *Cities* **2007**, *24*, 285–297. [CrossRef]

70. Hartig, T.; Evans, G.W.; Jamner, L.D.; Davis, D.S.; Gärling, T. Tracking restoration in natural and urban field settings. *J. Environ. Psychol.* **2003**, *23*, 109–123. [CrossRef]

71. Hausmann, A.; Slotow, R.O.B.; Burns, J.K.; Di Minin, E. The ecosystem service of sense of place: Benefits for human well-being and biodiversity conservation. *Environ. Conserv.* **2016**, *43*, 117–127. [CrossRef]

72. Hillsdon, M.; Panter, J.; Foster, C.; Jones, A. The relationship between access and quality of urban green space with population physical activity. *Public Health* **2006**, *120*, 1127–1132. [CrossRef] [PubMed]

73. Hock Teck, L.H.; Chin Siong, H.; Ali, H.M.; Tu, F. Do institutions matter in neighbourhood commons governance? A two-stage relationship between diverse property-rights structure and residential public open space (POS) quality: Kota Kinabalu and Penampang, Sabah, Malaysia. *Int. J. Commons* **2016**, *10*, 294–333. [CrossRef]

74. Horan, E.; Craven, J.; Goulding, R. Sustainable urban development and liveability. How can Melbourne retain its title as the world's most liveable city and strive for sustainability at the same time? *Eur. J. Sustain. Dev.* **2014**, *3*, 61–70. [CrossRef]

75. Howley, P.; Scott, M.; Redmond, D. Sustainability versus liveability: An investigation of neighbourhood satisfaction. *J. Environ. Plan. Manag.* **2009**, *52*, 847–864. [CrossRef]

76. Ikin, K.; Le Roux, D.S.; Rayner, L.; Villaseñor, N.R.; Eyles, K.; Gibbons, P.; Manning, A.D.; Lindenmayer, D.B. Key lessons for achieving biodiversity-sensitive cities and towns. *Ecol. Manag. Restor.* **2015**, *16*, 206–214. [CrossRef]

77. Irvine, K.N.; Devine-Wright, P.; Payne, S.R.; Fuller, R.A.; Painter, B.; Gaston, K.J. Green space, soundscape and urban sustainability: An interdisciplinary, empirical study. *Local Environ.* **2009**, *14*, 155–172. [CrossRef]

78. Kaźmierczak, A. The contribution of local parks to neighbourhood social ties. *Landsc. Urban Plan.* **2013**, *109*, 31–44. [CrossRef]

79. Kurniawati, W. Public space for marginal people. *Procedia-Soc. Behav. Sci.* **2012**, *36*, 476–484. [CrossRef]

80. Malek, N.A.; Mariapan, M.; Ab Rahman, N.I.A. Community participation in quality assessment for green open spaces in Malaysia. *Procedia-Soc. Behav. Sci.* **2015**, *168*, 219–228. [CrossRef]

81. Manfredo, M.J.; Teel, T.L.; Dietsch, A.M. Implications of human value shift and persistence for biodiversity conservation. *Conserv. Biol.* **2016**, *30*, 287–296. [CrossRef]

82. Massey, D. Liveable town and cities: Approaches for planners. *Town Plan. Rev.* **2005**, *76*, 1–6. [CrossRef]

83. Nasution, A.D.; Zahrah, W. Public open space privatization and quality of life, case study Merdeka Square Medan. *Procedia-Soc. Behav. Sci.* **2012**, *36*, 466–475. [CrossRef]

84. Revell, G.; Anda, M. Sustainable urban biophilia: The case of greenskins for urban density. *Sustainability* **2014**, *6*, 5423–5438. [CrossRef]

85. Schipperijn, J.; Ekholmb, O.; Stigsdottera, U.K.; Toftagerb, M.; Bentsena, P.; Kamper-Jørgensenb, F.; Randrupa, T.B. Factors influencing the use of green space: Results from a Danish national representative survey. *Landsc. Urban Plan.* **2009**, *95*, 130–137. [CrossRef]

86. Schneider, J.; Lorencová, H. Recreational activities, practices and attitudes of visitors to the protected landscape areas as a basis for resolving conflicts of recreation and nature protection. *Acta Univ. Agric. Silvic. Mendel. Brun.* **2015**, *63*, 1555–1564. [CrossRef]

87. Shackleton, S.; Chinyimba, A.; Hebinck, P.; Shackleton, C.; Kaoma, H. Multiple benefits and values of trees in urban landscapes in two towns in northern South Africa. *Landsc. Urban Plan.* **2015**, *136*, 76–86. [CrossRef]

88. Shamsuddin, S.; Hassan, N.R.A.; Bilyamin, S.F.I. Walkable environment in increasing the liveability of a city. *Procedia-Soc. Behav. Sci.* **2012**, *50*, 167–178. [CrossRef]

89. Shanahan, D.F.; Lin, B.B.; Bush, R.; Gaston, K.J.; Dean, J.H.; Barber, E.; Fuller, R.A. Toward improved public health outcomes from urban nature. *Am. J. Public Health* **2015**, *105*, 470–477. [CrossRef]

90. Shanahan, D.F.; Lin, B.B.; Gaston, K.J.; Bush, R.; Fuller, R.A. What is the role of trees and remnant vegetation in attracting people to urban parks? *Landsc. Ecol.* **2015**, *30*, 153–165. [CrossRef]

91. Soga, M.; Yamaura, Y.; Aikoh, T.; Shoji, Y.; Kubo, T.; Gaston, K.J. Reducing the extinction of experience: Association between urban form and recreational use of public greenspace. *Landsc. Urban Plan.* **2015**, *143*, 69–75. [CrossRef]

92. Staats, H.; Kieviet, A.; Hartig, T. Where to recover from attentional fatigue: An expectancy-value analysis of environmental preference. *J. Environ. Psychol.* **2003**, *23*, 147–157. [CrossRef]

93. Stanley, M.C.; Beggs, J.R.; Bassett, I.E.; Burns, B.R.; Dirks, K.N.; Jones, D.N.; Linklater, W.L.; Macinnis-Ng, C.; Simcock, R.; Trowsdale, S.A.; et al. Emerging threats in urban ecosystems: A horizon scanning exercise. *Front. Ecol. Environ.* **2015**, *13*, 553–560. [CrossRef]

94. Sugiyama, T.; Gunn, L.D.; Christian, H.; Francis, J.; Foster, S.; Hooper, P.; Owen, N.; Giles-Corti, B. Quality of public open spaces and recreational walking. *Am. J. Public Health* **2015**, *105*, 2490–2495. [CrossRef] [PubMed]

95. Sushinsky, J.R.; Rhodes, J.R.; Possingham, H.P.; Gill, T.K.; Fuller, R.A. How should we grow cities to minimize their biodiversity impacts? *Global Chang. Biol.* **2012**, *19*, 401–410. [CrossRef] [PubMed]

96. Taylor, B.T.; Fernando, P.; Bauman, A.E.; Williamson, A.; Craig, J.C.; Redman, S. Measuring the quality of public open space using Google Earth. *Am. J. Prev. Med.* **2011**, *40*, 105–112. [CrossRef] [PubMed]
97. Tonge, J.; Moore, S.A. Importance-satisfaction analysis for marine-park hinterlands: A Western Australian case study. *Tour. Manag.* **2007**, *28*, 768–776. [CrossRef]
98. Turner, W.R.; Nakamura, T.; Dinetti, M. Global urbanization and the separation of humans from nature. *BioScience* **2004**, *54*, 585–590. [CrossRef]
99. Van Herzele, A.; Wiedemann, T. A monitoring tool for the provision of accessible and attractive urban green spaces. *Landsc. Urban Plan.* **2003**, *63*, 109–126. [CrossRef]
100. Villanueva, K.; Badland, H.; Hooper, P.; Koohsari, M.J.; Mavoa, S.; Davern, M.; Roberts, R.; Goldfeld, S.; Giles-Corti, B. Developing indicators of public open space to promote health and wellbeing in communities. *Appl. Geogr.* **2015**, *57*, 112–119. [CrossRef]
101. Wetzstein, S. *Perceptions of Urban Elites on Four Australasian Cities: How Does Perth Compare?* Committee for Perth, University of Western Australia: Perth, Australia, 2010; pp. 1–17.
102. Zhang, W. Research on how to Improve the Liveability of City Community. *Appl. Mech. Mater.* **2012**, *174–177*, 3503–3506. [CrossRef]
103. Allen, N. Understanding the importance of urban amenities: A case study from Auckland. *Buildings* **2015**, *5*, 85–99. [CrossRef]
104. Child, S.T.; McKenzie, T.L.; Arrendondo, E.M.; Elder, J.P.; Martinez, S.M.; Ayala, G.X. Associations between park facilities, user demographics, and physical activity levels at San Diego County parks. *J. Park Recreat. Adm.* **2014**, *2014 32*, 68–81.
105. Lin, B.B.; Fuller, R.A.; Bush, R.; Gatson, K.J.; Shanahan, D.F. Opportunity or orientation? Who uses urban parks and why. *PLoS ONE* **2014**, *9*. [CrossRef] [PubMed]
106. Johnson, A.J.; Glover, T.D. Understanding urban public space in a leisure context. *Leis. Sci.* **2013**, *35*, 190–197. [CrossRef]
107. Matsuoka, R.H.; Kaplin, R. People needs in the urban landscape: Analysis of landscape and urban planning contributions. *Landsc. Urban Plan.* **2008**, *84*, 7–19. [CrossRef]
108. Cracknell, D.; White, M.P.; Pahl, S.; Nichols, W.J.; Depledge, M.H. Marine biota and psychological well-being: A preliminary examination of dose-response effects in an aquarium setting. *Environ. Behav.* **2016**, *48*, 1242–1269. [CrossRef] [PubMed]
109. Davern, M.; Farrar, A.; Kendal, D.; Giles-Corti, B. *Quality Green Space Supporting Health, Wellbeing and Biodiversity: A Literature Review*; National Heart Foundation of Australia: Adelaide, Australia, 2017; pp. 1–39, ISBN 978-0-9872841-7-4.
110. Gladwell, V.F.; Brown, D.K.; Wood, C.; Sandercock, G.R.; Barton, J.L. The great outdoors: How a green exercise environment can benefit all. *Extreme Physiol. Med.* **2013**, *2*, 1–7. [CrossRef] [PubMed]
111. Li, Q.T.; Otsuka, M.; Kobayashi, Y.; Wakayama, H.; Inagaki, M.; Katsumata, Y.; Hirata, Y.; Li, Y.; Hirata, K.; Shimizu, T.; et al. Acute effects of walking in forest environments on cardiovascular and metabolic parameters. *Eur. J. Appl. Physiol.* **2011**, *111*, 2845–2853. [CrossRef] [PubMed]
112. Rupprecht, C. Informal urban green space: Residents' perception, use, and management preferences across four major Japanese shrinking cities. *Land* **2017**, *6*, 59. [CrossRef]
113. Specht, R.L. Vegetation. In *The Australian Environment*, 4th ed.; Leeper, G.W., Ed.; CSIRO-Melbourne University Press: Melbourne, Australia, 1970; pp. 44–67.
114. Onnom, W.; Tripathi, N.; Nitivattananon, V.; Ninsawat, S. Development of a liveability city index (LCI) using multi criteria geospatial modelling for medium class cities in developing countries. *Sustainability* **2018**, *10*, 520–539. [CrossRef]

 *land*

Article

Environmental Justice in Accessibility to Green Infrastructure in Two European Cities

Catarina de Sousa Silva [1], Inês Viegas [1], Thomas Panagopoulos [1,*] and Simon Bell [2]

[1] Research Centre of Tourism, Sustainability and Well-being, University of Algarve, Gambelas Campus, 8005 Faro, Portugal; a47403@ualg.pt (C.d.S.S.); a62639@ualg.pt (I.V.)
[2] Chair of Landscape Architecture, Institute of Agricultural and Environmental Sciences—Estonian University of Life Sciences, Kreutzwaldi 5, 51006 Tartu, Estonia; simon.bell@emu.ee
* Correspondence: tpanago@ualg.pt; Tel: +351-289800900

Received: 20 October 2018; Accepted: 8 November 2018; Published: 12 November 2018

Abstract: Although it is well-established that urban green infrastructure is essential to improve the population's wellbeing, in many developed countries, the availability of green spaces is limited or its distribution around the city is uneven. Some minority groups may have less access or are deprived of access to green spaces when compared with the rest of the population. The availability of public green spaces may also be directly related to the geographical location of the city within Europe. In addition, current planning for urban regeneration and the creation of new high-quality recreational public green spaces sometimes results in projects that reinforce the paradox of green gentrification. The aim of this study was to explore the concept of environmental justice in the distribution of the public green spaces in two contrasting cities, Tartu, Estonia; and Faro, Portugal. Quantitative indicators of public green space were calculated in districts in each city. The accessibility of those spaces was measured using the "walkability" distance and grid methods. The results revealed that there was more availability and accessibility to public green spaces in Tartu than in Faro. However, inequalities were observed in Soviet-era housing block districts in Tartu, where most of the Russian minority live, while Roma communities in Faro were located in districts without access to public green space. The availability of public green spaces varied from 1.22 to 31.44 m^2/inhabitant in the districts of Faro, and 1.04 to 164.07 m^2/inhabitant in the districts of Tartu. In both cities, 45% of the inhabitants had accessible public green spaces within 500 m of their residence. The development of targeted new green infrastructure could increase access to 88% of the population for the city of Faro and 86% for Tartu, delivering environmental justice without provoking green gentrification. The outcome of this study provides advice to urban planners on how to balance green space distribution within city neighbourhoods.

Keywords: urban sustainability; landscape urbanism; green gentrification; Soviet-era housing blocks; deprived areas; Roma minority

1. Introduction

Urban green infrastructure (UGI), which includes parks, community gardens, forests, and corridors along waterways, provides important connections between communities and nature [1]. Urban green infrastructure planning aims to develop green space networks in limited spaces in compact cities [2]. These areas provide numerous ecosystem services, and have contributed to increasing physical activity, improvement in mental health, and improved socialisation of the community residents [3]. Urban green infrastructure helps cities increase their resilience to climate change and to improve their attractiveness by offering a cleaner and healthier environment [4].

Environmental justice occurs when green infrastructure is equally distributed, without discrimination, within a city. Cities are complex social-ecological systems, and the decisions and processes responsible for today's UGI availability have occurred over centuries; for this reason, it is difficult to determine all the factors that affect the quality of urban green infrastructure. For example, along with the environmental justice issues, the location of a city, related to morphological and climatic conditions, such as water availability, proves to be another strong characteristic that affects the availability of urban green spaces. Although it is well-established that the relationship between green infrastructure and the urban environment is essential for improving wellbeing and population health [3], in many developed countries, the availability of green space is limited, or its distribution across the city is uneven.

Some authors [5,6] consider that the availability of green spaces in European cities may be directly related to their geographical location, with higher amount of public green spaces found in Northern and Central European countries than in Mediterranean ones. Furthermore, it is also recognised that the distribution of public green space is related to the location of different social classes [7]. This means that socioeconomically deprived sections of society, such as low-income groups or ethnic minorities, often also have less access or are deprived of access to green spaces, compared to the rest of the population [8].

Low accessibility to UGI in some areas of a city or for some demographic groups is, nowadays, a problem of environmental justice. For this reason, projects have been developed with the aim of rehabilitating degraded neighbourhoods and increasing the amount of green space in order to improve wellbeing and environmental justice. The implementation of those projects has been successful when they are integrated into the existing green infrastructure network and designed according to the needs of the inhabitants. Conversely, paradoxical interventions have resulted in the phenomenon of green gentrification.

Green gentrification has been observed when urban regeneration projects attract investment around new, high-quality public green space, and become an attraction for social classes with higher income and greater purchasing power. The demand for accommodation near these new recreational spaces promotes the renovation of dwellings or the construction of new ones and, therefore, the house process, rental values, and living costs increase. Consequently, the original residents are forced to relocate to cheaper areas of the city, but they lose the benefits of the public green spaces that were provided to improve their wellbeing. Green gentrification is, however, a difficult phenomenon to study since it is only visible in the long term [9].

Estonia is one of the greenest countries in Europe [5], but there are green infrastructure differences between the Soviet-era housing districts and the districts developed both before and after that period (1944–1991). Research into the inequalities of post-Soviet Union has focused on the quality of buildings, but ignored the accessibility of public green spaces. Generally, accessibility to public green spaces is lower in Mediterranean cities compared with North European ones [10,11]. Faro is one of the Portuguese cities with the least green space per inhabitant (8 m^2) [9,12]. Thus, there are two pertinent reasons for comparing the accessibility of public green space in Tartu, Estonia; and Faro, Portugal. Both cities are similar in terms of population size (small cities) and have similar social and cultural dynamics, however, they are distinct geographically and morphologically.

This article explores the notion of environmental justice in urban green infrastructure and how to avoid green gentrification. The overall aim of the study was to investigate the accessibility of public green spaces in a Nordic/Baltic and a Mediterranean city, and addressed the following research questions: Are there districts with more and less accessibility in Tartu and Faro? Are the minority groups and time of construction related to the accessibility of the public green spaces? The hypothesis of this study was that "there is no environmental justice in the distribution of the public green spaces in both Tartu and Faro".

2. Literature Review

2.1. Green Infrastructure and Ecosystem Services

There are different perspectives on the definition of green infrastructure components. One model of green infrastructure is a system of hubs, links, and sites [13]. Hubs are considered to be the anchor of the green infrastructure network, since they provide space for native plants and animal communities, habitat for wildlife and people, and are responsible for the ecological processes that move through the system. They are the most ecologically important large natural areas, the habitat of many species, and are sometimes essential to support particular life stages of several species. Links or corridors are linear elements responsible for connecting hubs to allow the flux of animals and plants. They are essential for preventing the extinction of (many) species [14]. Finally, Ahern [15] consider that the Green Infrastructure concept assumes the same mosaic model used in landscape ecology, which categorises three major landscape elements: patches, corridors, and matrix.

In the end of 19th and early 20th century, the landscape architect Frederick Law Olmsted stated that all urban green areas, independent of their characteristics, should provide people with benefits from nature. For this reason, he considered that parks should be connected to each other and to surrounding residential areas [16]. These two ideas were in the origin of the greenway movement that, by the end of the 20th century, would evolve into the term "green infrastructure" [13]. There are two concepts that formed the origin of this idea: (1) connecting all green spaces for the benefit of citizens, (2) preserving and linking natural areas to counter habitat fragmentation and promote biodiversity. These two concepts are very similar to the ideas developed by Olmsted and implemented in the 1880s in the revolutionary Emerald Necklace in Boston [13].

A commonly used definition describes green infrastructure "as a strategically planned network of high-quality natural and semi-natural areas with other environmental features, which is designed and managed to deliver a wide range of ecosystem services and protect biodiversity in both rural and urban settings" [17].

There are four classes of ecosystem services [18,19]: (1) provisioning services (for example, food, wood, fibre); (2) regulating services (for example, air quality regulation, climate regulation, water regulation); (3) cultural services, and (4) supporting services. Cultural services include all the non-material benefits that people obtain from ecosystems, such as spiritual enrichment, cognitive development, reflection, recreation, and aesthetic experiences, together with other important social benefits, like the improvement of mental health [20], stress reduction, and relaxation [21]. Furthermore, urban green infrastructure can act as a meeting place in a neighbourhood and positively influence the interactions between different communities [22]. Finally, the supporting services include soil formation, photosynthesis, nutrient cycling, and water cycling. These include all the services that are necessary for the production of all other ecosystem services [23].

2.2. Environmental Justice

Although there is agreement on the importance of urban green infrastructure and that this should be based on the three pillars of sustainability—ecology, economy, and society—many studies [24,25] reveal that the societal variables are least respected, or even ignored, in the process of project development. The exclusion of one sustainability variable can results in an unequal distribution of green infrastructure through the city. This uneven distribution is often stratified based on socioeconomic or ethno-racial characteristics, including age, gender, disability, education and wealth of the residents, and has been recognised in the literature as an environmental justice issue [26,27]. A definition of environmental justice is the right of the entire population to be protected against environmental pollution and to live in a clean and healthful environment [28].

Traditionally, the environmental justice movement has focused on pollution issues affecting the health of low-income and minority individuals who lived in close proximity to polluting sites [29,30]. The movement appeared in the beginning of the 1980s in the United States, where low-income ethnic communities, including African-Americans and Hispanics, were living in the most polluted neighbourhoods, compared to the white majority communities who, conversely, were living in neighbourhoods of high environmental quality [31–33]. More recently, environmental justice research has focused on the distribution of environmental hazards and amenities, including green spaces, through different social groups [34–38].

Gould and Lewis [28] refer the importance of analysing the full spectrum of distribution, meaning that research should also consider who gets access to the environmental amenities, such as parks, water clean-up, and access to public transport, by studying all social groups, instead of only those that bear the environmental burdens of society, such as toxic waste, hazardous facilities, and poor air quality, which has been the focus up to now.

One of the most common ways to measure inequality among the population is through the Gini index, which is used in economics. However, a single index to measure environmental inequality does not yet exist [39]. Thus, in making cities more resilient, equitable, and sustainable, it is necessary to be aware of environmental justice problems, and to emphasise the questions of access to urban green spaces [24].

2.3. Accessibility to Public Green Spaces

Public urban green space is defined as public parks and other green spaces that are accessible to the general public and managed by the local government [40]. Lindholst et al. [41] define three main characteristics to evaluate the quality of urban green spaces: (1) structure and general aspects, such as size, character, location, and accessibility; (2) functionality and experience, for example, the recreational and social aspects, culture and history, nature and biodiversity, landscape and aesthetics, and environment and climate; and (3) management and organisation, including management, maintenance, and communication and information.

There are inequalities in the distribution of green infrastructure in most cities: the different groups of society have more or less access to green infrastructure, depending on their socioeconomic status (SES). In order to study the inequalities of access to UGI and to provide solutions, it is necessary to measure it. Most research on accessibility has focused primarily on two aspects: (1) distance to green spaces and (2) the area available at that distance, providing threshold values of urban green space per habitant. However, in some cases, accessibility has been estimated using only one of these factors [42]. The European Environment Agency (EEA) recommends that people should live within 15 min walking distance of their place of residence [43], but does not specify the available area of green space per resident. Also, Wolch et al. [35] defined 400 m, a five-minute trip, as the standard distance between a public park and people's house. In other studies, both aspects have been combined. Coles and Bussey [44] considered that green spaces should be a 5- to 10-min walk from the residence area, and have a minimum area of 2 ha. Van Herzele and Wiedemann [45] suggested a 5-min walk, equivalent to 400 m, to the closest 1–10 ha green space. The UK government agency, English Nature, recommends, in the Accessible Natural Greenspace Standards, that at least 2 ha of accessible natural green space should be provided per 1000 population, with a minimum distance of 300 m from the place of residence [46,47]. Magalhães [48] considers a minimum distance for children and elderly people of 100 m, and also considers 400 m^2 as the minimum area for a public green space in Portugal. The World Health Organization [49] assumes a minimum of 9 m^2 green space per person, and the ideal minimum area of green space should be 50 m^2.

According to Maroko et al. [50], accessibility to public urban green spaces can be measured with the container approach, the walkability distance method, and the Kernel density estimation. The container approach measures the accessibility using a particular geographic unit of aggregation, such as zip code, neighbourhood, or census unit, to determinate the location of a park or recreational facility, instead of using a proximity measure. In this method, the number of parks per areal unit can be estimated for the unit of aggregation used, and related to specific populations characteristics, for example, SES [51,52]. The walkability distance method considers a standard walking distance

(5–10 min walk, 400 m or 800 m) to parks as a proxy for access. Nevertheless, in this method, the actual street network was not considered, only Euclidean distance. Meanwhile, the relationship between distance and willingness to walk is a continuous curve without sharp breaks, thus, the Kernel density estimation used by Moore et al. [53] may estimate, more accurately, the accessibility for every point of a study area, because it uses blocks of areas, instead of giving a binary answer of accessible or not in just a few metres of distance.

According to Fan et al. [40], five variables should be included when evaluating access to public urban green spaces: (1) a citizen-based opinion, reflecting the quality of a green space where residents live; (2) multiple functional levels, including a quantitative evaluation of the green space from neighbourhood to city level according to their functional scales; (3) preconditions for users, for example, accessibility and safety; (4) a quality measure that assesses the variety of suitability of green spaces to accommodate different activities; and (5) multiple uses according to the diverse conditions [45]. Meanwhile, Dai [54] argues that a common descriptive approach is based on the availability of green spaces per inhabitant, calculating the rate of the supply vs the demand within a predefined region. However, it is not completely always predictable that people go to the closest green space for various reasons, such as its size, fear of dogs, or fear of crime and racial attacks.

2.4. Gentrification and Green Gentrification

Gentrification is defined by Smith [55] as "the process by which working class residential neighbourhoods are rehabilitated by middle class homebuyers, landlords, and professional developers"; moreover, this process emerged as a sporadic and local anomaly in the housing markets of some large cities. Most recently, Smith [56] assume that this process is currently generalised as an urban approach and a neoliberal urban policy, and gives the example of the working-class quarters in London that, gradually, were replaced by expensive residences. This process of gentrification, once installed completely, changed the original social character and expanded rapidly to other districts with similar characteristics. According to [56], in Western Europe and the United States, there were three historical waves in the gentrification process, beginning in the 1950s as sporadic gentrification in small neighbourhoods; the second wave, during the 1970s and 1980s, related to urban and economic restructuring processes; and the third wave, which emerged in the 1990s, defined as the recessional pause and subsequent expansion.

Gould and Lewis [57] describe green gentrification as the process of displacement or exclusion of the economically most-vulnerable classes of society, which is enabled by the creation or renovation of an environmental amenity [58]. Currently, urban regeneration projects in degraded areas have been promoted as improving the wellbeing of residents and solving environmental injustice problems. However, such environmental improvements in ethnic communities and/or low-income households can create an urban green space paradox, as already noted [35]. The creation of new, high-quality green spaces can increase attractiveness, making these neighbourhood more desirable. By contrast, the cost of housing can rise, and residents may not be able to afford the rent. This results in the exclusion or displacement of the poor neighbourhood's residents, who were intended to benefit from the ecosystem services provided by the new green space. In turn, the residents may only be able to afford to live in a similar degraded neighbourhood to the one they left, with low access to green infrastructure [59]. Such a phenomenon has been variously termed as ecological gentrification [58], eco-gentrification, green gentrification [57], or environmental gentrification [60].

One of the most famous examples of an intervention in an obsolete infrastructure that resulted in green gentrification is The High Line, in New York, designed by James Corner. The High Line is a linear park constructed on an abandoned elevated railway that was originally designed to facilitate access to factories and other businesses. This project has now become one of the most popular parks in New York City, attracting millions of visitors each year. Nevertheless, what appeared to be a successful project resulted in a case of green gentrification. The older and typically low-income industrial houses were rehabilitated, making them more liveable and attractive. This caused the displacement and exclusion of the residents who were unable to pay the rent on the rehabilitated properties, which led to their being rehoused in other degraded neighbourhoods [35].

Another much earlier example of green gentrification was Prospect Park in Brooklyn, designed by Frederick Law Olmsted and Calvert Vaux. Alongside the park, there was a clear difference between the ethnicities that occupied the richest neighbourhoods and the poorest ones. Over the years, the greening of the richest neighbourhood increased the value of houses and the opposite affected the degraded neighbourhood, so "the richest neighbourhood became richer, the poorest neighbourhood became poorer" [57]. Recently, with the restoration of Prospect Park, the rents of all neighbouring houses increased, resulting in the displacement of the poorest residents of the Prospect Park neighbourhood, and, consequently, in a renewed phenomenon of green gentrification [57].

However, sometimes the study of green gentrification can be hampered because this phenomenon could take a long period of time to appear—as parks take time to mature—and because the green gentrification phenomenon is directly related with the social inequalities, for this it can be difficult to study in cities that avoid to deal with the problem of social justice [61].

To control the effects caused by green gentrification, Curran and Hamilton [26] suggest a "just green enough" strategy, which consists of securing the public health benefits of enhanced access to urban green infrastructure while avoiding the urban green space paradox by promoting small-scale interventions in scattered sites. For example, Wolch et al. [35] promote urban allotments instead, big-scale projects that radically change the dynamics of these communities. Vancouver, British Columbia, and Michigan, USA, have significant and growing urban agriculture movements that adopt urban agriculture as a sustainability fix [62]. Another example is the grassroots movements for urban agriculture that has become popular in Detroit among the declining heavy industry, abandoned buildings, and shrinking population [63]. Another solution to control green gentrification is to involve minorities in the decision-making and planning for green spaces, to include their ethnic and cultural customs and perceptions [64].

2.5. The Socialist City

The socialist city was defined by Demko and Regulska [65] as the one that "No social or occupational group would have better or more favourably located residential sites. Similarly, public services of all kinds, including transportation, should be of equal quality, availability, and accessibility. Such amenities as a high-quality physical environment, including recreational environment, would be equally accessible to all. All such urban conditions would be similarly equitably arranged and available" (p. 290). In short, the socialist city had, as its main principle, an egalitarian society with equitable distribution and accessibility of amenities. However, the spatial patterns of inequality in both under capitalism and socialism periods have distinct interpretations and, for that reason, it is not appropriate to make strong comparisons [66].

The residential districts or microregions found in Soviet-era cities, providing mass-housing with basic consumer services, were known as the *mikrorajon*, and represented the "basic building block of the Soviet city" [66] (p. 75). These are very common in Estonian cities, for example, in Tartu, the second biggest city of Estonia. Constructed from the late 1950s through to the 1980s, most of the new housing consisted of Soviet-type brick or concrete panel dwellings [67]. In Estonian cities, it is possible to identify a trend in housing occupancy related to the different ethnic groups [68]. In general, Estonians live in single-family houses, while Russians (or Russian speakers—they may also have Belorussian or Ukrainian, etc., origins) occupy the high-density flats built during the Soviet Union (Figure 1). However, Kulu [69] admits that, for example, in late-Soviet Tartu, in spite of the fact that Estonians generally had more living space than non-Estonians, in the case of facilities, the situation was the opposite—non-Estonians had more facilities, even when housing ownership was controlled. These differences were due to the differences in preference for housing between ethnic groups, and also by the Soviet policy for social housing [68]. However, Kährik and Tammaru [70] admit that these neighbourhoods have still a "strong social mix, and do not reveal clear signs of decline" (p. 215).

2.6. Landscape Urbanism and Smart Growth

In order to deal with the environmental justice issues, new planning and development solutions should be adopted. Amongst the landscape architecture community, it is commonly accepted that the needs and perceptions of the people who use cities must be included in urban design in a co-creation process [71]. New urbanism, landscape urbanism, and smart growth are recent approaches, with a sustainability base that focuses on a human-scaled urban design.

New urbanism is a planning and development model based on the principles of walkability, housing, and shopping being in close proximity, and accessible public spaces. This aims to offer alternatives to the sprawling, single-use, low-density patterns, which have been shown to inflict negative economic, health, and environmental impacts on communities [72]. New urbanism replaces the large-scaled planning based on automobile circulation, and can be applied to diverse scales of development, for example, suburban areas, urban neighbourhoods, dense city centres, or even a single street. The field of the projects include new development, urban infill, and revitalisation or preservation design. From the late 1990s, the phrase "landscape urbanism" started to be used by landscape architects in the United States to refer to the redevelopment concepts for declining post-industrial cities [73]. Landscape urbanism is a mode of urban planning arguing that the best way to organise a city is through the design of its landscape, rather than the design of its buildings [74].

At the same time, the smart growth model is a development approach that encourages a mix of building types and uses, diverse housing and transportation options, development within an existing neighbourhood, and community engagement [75,76]. Smart Growth [77] define ten principles as supportive of the smart growth approach: (1) mixed land uses; (2) compact design; (3) range of housing choices; (4) walkable neighbourhoods; (5) communities with a strong sense of place; (6) preservation of open space, farmland, natural beauty, and critical environmental areas; (7) direct development towards existing communities; (8) providing a variety of transportation choices; (9) making fair and cost-effective development decisions; and (10) encouraging community and stakeholder collaboration in development decisions.

3. Materials and Methods

3.1. Study Area

The data used in this paper were gathered from Tartu City, Estonia; and Faro, Portugal (Figure 2). Estonia is a Northern European country located on the eastern coast of the Baltic Sea. It is a small country of approximately 45,227 km^2 in area, and around 1.3 million habitants, and is one of the most forested countries in the Europe Union [37]. During the Soviet period, between 1944 to 1991, the majority of the Estonia's cities, including Tallinn and Tartu, developed according to the principles of the socialist city, as discussed above. The urban planning and construction system during the

Soviet period favoured the development of large homogeneous areas, composed of extensive areas of multi-storey housing blocks [67].

Figure 2. Location of Tartu, Estonia; and Faro, Portugal.

Tartu is the second biggest city of Estonia, with 38.58 km^2 of total area (3858 ha urban area). According to the 2011 census, Tartu city has some 97,600 inhabitants, including 81.7% Estonians, 14.7% Russians, and 3.6% of other national origins, for example Ukrainians, Belarussians, or Finns. For this reason, Tartu is considered a socially mixed city [70]. It is known as a historic university city, while the industrial and military investments during the Soviet Union made it an important migrant destination [78]. Tartu is organised into 17 districts, and has twenty public green spaces with at least one hectare being of high quality. The most popular public green spaces in Tartu are Keskpark, Botaanikaaed (Botanic Garden)—with a wide variety of exotic species, Holmi park, Anne Kanali park (located along the Emajõgi River), and Toomemägi (Figure 3).

Faro is a southern Portuguese municipality and the capital of the Algarve region, with around 65,000 habitants (around 47,000 living in the city area). The city is about 711 ha and is bounded by the "Ria Formosa" Natural Park, which offers a picturesque landscape setting to the city. To increase the complexity of the analyses and to allow comparison of Faro with Tartu, eight districts were defined: Old Town; Urban centre; Alto Rodes; São Luís; Bom João and Industrial area; Alto de Santo António; and Penha and Figuras, based on the morphological characteristics of the city, the road network, time of construction, and the sections defined by the National Institute of Statistics. The Roma minority is located mainly in the districts of Penha, Industrial area, Figuras, and the urban periphery.

There are four public green spaces in Faro of at least 1 ha in area. Alameda garden is the oldest public green space of the city (Figure 3), with several facilities, including a playground, a miniature golf course, and senior citizen outdoor fitness equipment. Mata do Liceu is a public park used mainly for the practice of physical exercise. Parque Ribeirinho establishes a visual connection with Ria Formosa and, with about 16 ha, is the biggest and most recent public green space in the city. Parque Lazer is a small and flexible green used mainly for sports practice and to host small events or fairs. Arranged in a minimalistic way with a large (0.85 ha) central open field, the rest of the space is arranged peripherally with places of activity which continue to be developed with the help of some recurring local initiatives [9].

Figure 3. Toomemägi in Tartu (a) and Alameda garden in Faro (b). Image from field survey by authors.

3.2. Indicators of Availability and Accessibility to Urban Green Infrastructure

Following Dai [54], the area of public green spaces (PGS) was measured in each district using a geographic information system (GIS). The availability of PGS per inhabitant of each city district was calculated using data from the 2011 census. The relation, PGS per inhabitant, resulted in two social sustainability indicators that allowed us to compare them to other cities' values, and to clarify the degree of environmental justice in the distribution of green infrastructure through the city's districts. In Tartu, it was especially pertinent to compare the area of PGS per inhabitant in districts developed during the Soviet period with the newer ones.

The accessibility of PGS was measured using the walkability distance method, with buffers around the green spaces and within the administrative boundaries of 300 m (4-min walking distance) and 500 m (approximately 7-min walking distance). There are several variations in the literature regarding the minimum distance that a public green space should be from the place of residence [35,45,48]. For that reason, we considered it appropriate to use the two buffer distances and make a comparison of the results.

In order to compare the accessibility results in Tartu and Faro, we included public green spaces with of least 1 ha area, even if similar studies only consider the spaces of more than 2 ha [6]. This was because public green spaces in Mediterranean cities tend to be smaller than the North European ones.

The Estonian Land Board, Green Map System, Tartu Linn open data, and Google Earth were the sources of data to identify the current public green spaces of high quality. The proposal for future public green spaces in Tartu was identified from the "Tartu Linna Üldplaneering", a general plan covering the period up to 2030. The public green spaces from Faro city were defined using a map provided by the municipal authority. For both cities, the individual and total area of public green spaces was measured using QGIS 2.16.3 (Geographic Information System. Open Source Geospatial Foundation Project).

To relate the population density and accessibility to the public green spaces of Tartu and Faro, we used a method based on that of Kabisch et al. [6]. The walkability distance method maps were developed using a buffer of 300 m and 500 m, and then intersected with a 1 ha grid within the city borders. In each grid unit, the area of public green spaces per inhabitant was calculated. For each district, a different value of population density corresponding to the division between the total number of inhabitants and the total area of each district was used. Using different population densities allowed results to get closer to the reality of each district [53]. Finally, a five-level scale to classify the accessible area for public green spaces, in square metres per inhabitant, using the two different distance buffers, was developed.

In order to analyse the evolution of land use in Tartu city, statistical data, from between 1998 and 2017, was used. This information was important because it allowed us to make a deeper characterisation of the city, and revealed the evolution of the public parks and green areas in the city that can reflect the importance given to this use. In order to evaluate the access to the pubic green spaces based on the ethnic differences of the society, data was collected from the 2011 Estonian census about the number of ethnic nationalities per district of Tartu. In Faro, the 2011 census and our own observations were used to estimate the population of each district and the location of the Roma minority.

4. Results

The PGS per inhabitant was calculated in each district for both cities. Table 1 presents the area of gardens, buildings, trees, and grassland per district of Tartu to characterise the relative relation of green space and urbanised area in each district. There were two main types of buildings in Tartu city that are directly related to the area of yards. For example, the districts of Tammelinn, Ihaste, and Karlova have a higher proportion of yards, and the majority of the buildings in these districts were single-family houses.

Table 2 presents the location of ethnic minorities in the city, and compares this with the distribution of public green spaces in the city. The districts with higher numbers of ethnic minorities, mainly Russians, were Jaamamõisa (53% of Estonians, 39% of Russians, and 8% of other nationalities), Annelinn (67% of Estonians, 28% of Russians, and 5% of other nationalities), and Ropka tööstusrajoon (80.1% of Estonians, 16.4% of Russians, and 3.4% of other nationalities). On the other side, the districts with fewer ethnic minorities were Variku, Tähtvere, Tammelinn, and Karlova.

Table 1. Area of land use per district of Tartu, Source: Estonian Land Board.

Districts	Area (ha)	Gardens (ha)	Buildings (ha)	Trees (ha)	Grassland (ha)	Other (ha)
Annelinn	542	41.3	31.67	43.6	225.0	113.6
Ihaste	425	129.7	21.90	99.8	118.7	8.7
Jaamamõisa	143	26.1	10.38	8.1	37.1	25.2
Karlova	229	118.8	42.85	0.8	11.3	24.0
Kesklinn	180	45.3	38.31	15.5	5.9	34.1
Maarjamõisa	135	39.5	13.42	13.4	15.9	19.7
Raadi-Kruusamäe	290	107.8	30.18	42.6	57.3	39.0
Ropka tööstusrajoon	360	122.2	36.92	2.7	115.2	38.5
Ropka	147	71.9	21.77	9.6	11.8	21.6
Ränilinn	122	9.8	9.55	4.2	17.5	13.0
Supilinn	70	30.9	7.27	1.8	19.7	0.5
Tammelinn	289	192.2	42.54	0.8	29.9	18.3
Tähtvere	228	99.0	21.30	18,5	69.2	40.1
Vaksali	77	58.0	14.72	0.0	9.3	3.8
Variku	77	37.5	9.18	0.0	20.9	2.4
Veeriku	280	121.6	35.65	8.9	51.8	31.6
Ülejõe	304	79.2	26.68	53.4	72.1	27.4

Table 2. Residents per ethnic nationality in Tartu districts. Source: Census, 2011.

District	Total Population	Estonian	Russian	Ukrainian	Belarusian	Other Nationalities
Annelinn	27,042	18,164	7560	468	177	676
Ihaste	2690	2326	293	20	8	72
Jaamamõisa	3399	1811	1314	106	47	121
Karlova	9627	8963	462	39	3	161
Kesklinn	6994	6045	584	26	14	336
Maarjamõisa	1454	1289	142	7	5	14
Raadi-Kruusamäe	4578	3783	659	38	20	80
Ropka tööstusrajoon	3247	2601	534	25	15	73
Ropka	5077	4656	333	25	5	65
Ränilinn	1678	1504	136	9	2	29
Supilinn	1925	1784	86	4	2	54
Tammelinn	6694	6356	237	20	6	77
Tähtvere	3434	3238	113	10	3	70
Vaksali	3126	2769	254	12	5	90
Variku	1773	1657	95	2	2	18
Veeriku	5411	4832	453	24	18	86
Ülejõe	9110	7695	1012	56	22	328

Table 3 presents the number of inhabitants and area of each district of Faro. The number of immigrants has been decreasing over the years. There were 11.2%, in 2010, and 9.9% by 2016. However, there is no available data relating ethnic nationality with place of residence. Despite this, it was possible to identify the Roma minority of Faro as being located in four main neighbourhoods: in Penha, in industrial area, and in the periphery. They live in social housing estate neighbourhoods, or in prefabricated houses and containers, built in areas far from the city centre and with difficult access to services, which seems to reinforce the segregation of the communities [79]. Furthermore, the area of the dwellings was very small, without adequate conditions for people with disabilities or elderly people, and without common or leisure spaces. These neighbourhoods have small entrances, and the composition forms a labyrinth, which makes it appear that the neighbourhood is "closed" to the rest of the city space [80]. There are still other Roma clusters of small dimensions, in Faro, that live in miserable conditions in abandoned old houses or containers.

Table 3. Area of districts in Faro and population, Source: Census 2011 and own calculations.

Districts	Total Area (ha)	Total Population	Roma Population	Density (inhab/ha)
Old Town and Historic centre	26	1170	-	45
Urban centre	52	3535	-	68
Alto Rodes	57	5659	<10	99
São Luís	58	7089	<10	122
Bom João and Industrial area	146	4141	>200	28
Alto de Santo António	105	6600	<10	63
Penha	100	8170	>100	82
Figuras and Urban Periphery	167	4851	>200	29

Table 4 for Tartu and Table 5 for Faro show the results of availability and accessibility to PGS of at least 1 ha in these cities. In Tartu, Tähtvere was the district with the greatest area of PGS per inhabitant, around 146 m^2 per habitant. The districts of Maarjamõisa and Supilinn also had a significant area of PGS per inhabitant, 126 m^2/inhabitant and 79 m^2/inhabitant, respectively. Conversely, Karlova was the district with the lowest area of PGS per habitant, about 1 m^2/inhabitant. Also, the districts of Ihaste, Jaamamõisa, Ropka, Ränilinn, Tammelinn Vaksali, Variku, and Veeriku had less than 10 m^2 per inhabitant. In Faro (Table 5), the majority of districts had less than 10 m^2 of PGS per inhabitant. The exception was district 8, "Figuras and Urban Periphery", with about 31 m^2/inhabitant.

Table 4. Availability and accessibility of public green spaces (PGS) in Tartu.

District	Area (ha)	Number of inhabitants	PGS (ha)	PGS Availability (m²/inhabitant)	Accessible Area in 300 m (ha)	Accessible in 300 m (%)	Accessible Area in 500 m (ha)	Accessible in 500 m (%)
Annelinn	542	27,042	39.89	14.75	176	32	280	52
Ihaste	425	2690	<1	3.72	16	4	46	11
Jaamamõisa	143	3399	<1	2.94	0	0	0	0
Karlova	229	9627	<1	1.04	73	32	134	59
Kesklinn	180	6994	29.21	41.76	150	83	173	96
Maarjamõisa	135	1454	18.34	126.13	105	78	133	99
Raadi-Kruusamäe	290	4578	16.05	35.06	72	25	117	40
Ropka tööstusrajoon	360	3247	7.26	22.36	53	15	95	26
Ropka	147	5077	2.97	5.85	28	19	49	33
Ränilinn	122	1678	<1	5.96	2	2	19	16
Supilinn	70	1925	15.17	78.81	70	100	70	100
Tammelinn	289	6694	1.17	1.75	68	24	124	43
Tähtvere	228	3434	56.34	164.07	164	72	213	93
Vaksali	77	3126	<1	3.20	39	51	53	69
Variku	77	1773	<1	5.64	0	0	0	0
Veeriku	280	5411	<1	1.85	17	6	47	17
Ülejõe	304	9110	8.90	9.77	120	39	194	64
Total	3898	97,259	202.3	20.8	1153	30	1747	45

Table 5. Availability and accessibility of public green spaces (PGS) in Faro.

District	Area (ha)	Number of inhabitants	PGS (ha)	PGS Availability (m²/inhabitant)	Accessible Area in 300 m (ha)	Accessible in 300 m (%)	Accessible Area in 500 m (ha)	Accessible in 500 m (%)
Historic centre	26	1170	<1	<8.55	0	0	6	23
Urban centre	52	3535	<1	<2.83	8	15	17	33
Alto Rodes	57	5659	<1	<1.77	19	33	39	68
São Luís	58	7089	<1	<1.41	19	33	31	53
Bom João & Industrial	146	4141	1.97	4.76	37	25	70	48
Alto Santo António	105	6600	3.59	5.44	41	39	74	70
Penha	100	8170	<1	<1.22	0	0	0	0
Figuras & Periphery	167	4851	15.25	31.44	59	35	85	51
Total	711	41,215	25.81	6.2	183	25	322	45

Figure 4 shows the map of accessibility to PGS from each residence, using the walkability method, for the city of Tartu, which was quantified in Table 4. It can be seen that using the 300 m and 500 m distance, and two different sources of data, 45% of the city residents have access to PGS within a seven-minute walk. Meanwhile, the same percentage (45%) was observed also for the city of Faro in the buffer of 500 m (Figure 5 and Table 5). This result shows that even though the South European city had much less available PGS per inhabitant, the accessibility was the same as in the very green, but more spread out, city of Northern Europe.

According to Table 4, it is possible to see that the districts with greater proportions of accessible area were Supilinn, where the accessible area covers the district's total area, for both distance buffers; Kesklinn (83%, using 300 m buffer and 96% using 500m buffer); Maarjamõisa (78% and 99%, respectively), and Tähtvere (72% and 93%, respectively). The districts with lower proportions of accessible area to PGS were Jaamamõisa and Variku, without accessible areas in either district; Ränilinn with only 2% considering a 300 m buffer and 16% considering a 500 m buffer; Veeriku with 6% and 16% respectively; and Ihaste with 4% and 11%, respectively. It is important to look at the percentage of each district, rather than the total accessible area of PGS, because it considers the total area of each district.

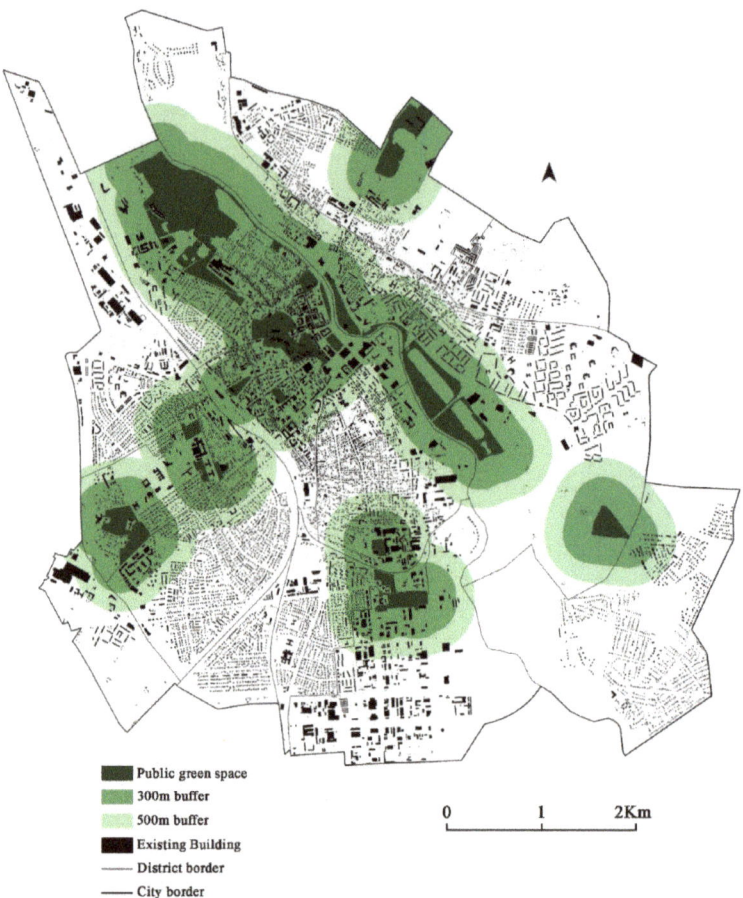

Figure 4. Accessibility of pubic green spaces (PGS) in Tartu, using the "walkability" distance method (300 and 500 m buffers).

Figure 5 shows the map of accessibility to PGS from each residence using the walkability method for the city of Faro, which was quantified in Table 5. It was revealed that the districts with higher proportions of accessible public green spaces were district 6 (Alto de Santo António) with 39%, using the 300 m buffer; and 70%, using the 500 m buffer; and district 8 (Figuras) with 35% and 51%, respectively, of the district's total area. On the other hand, the districts with lower proportions of accessible area were district 1 (Old Town and Historic centre) with only 23% using the 500 m buffer and district 7 (Penha) without any residence accessible within a 7-min walk to the PGS.

According to the grid method using the 300 m and 500 m buffers, the Tartu districts of Annelinn and Karlova had between 101 and 250 m^2 of accessible PGS per inhabitant, and the remaining districts had over 251 m^2 accessible PGS per inhabitant. Using the same method in Faro, it was revealed that, for the 300 m buffer, the districts 5 and 8 had more than 251 m^2 of accessible PGS per inhabitant, the districts 2 and 6 had between 101 and 250 m^2 per inhabitant, and the districts 3 and 4 had between 31 and 100 m^2 per inhabitant. The Old Town historic centre and Penha districts had no PGS. Considering the area of 500 m buffer, the districts 6 and 8 had more than 251 m^2 of accessible PGS, the districts 1, 2, 3, and 6 had between 101 and 250 m^2 of accessible PGS per inhabitant, and district 4 had 31–100 m^2 per inhabitant of accessible PGS. Penha was the only district without any accessible PGS, even within 500 m walking distance.

Figure 5. Accessibility of pubic green spaces (PGS) in Faro, using the "walkability" distance method (300 and 500 m buffer).

5. Discussion and Proposal for New Green Infrastructure

According to the evolution of the land use and occupation, it was confirmed that there was a significant increase in the public parks and green areas of Tartu, between 1998 and 2000. During the following years, since 2000 until 2017, the total area of public parks and green areas ranged between approximately 330 and 400 ha of total area, with their maximum value in 2016. In 2017, the total area of public parks and green areas (389 ha) corresponded to 10% of the city's total area. The total area of high-quality PGS was 202 ha. Meanwhile, in the present study, we considered only the high-quality PGS with more than 1 ha.

Analysing the data on ethnic minorities, it could be seen that the districts with the highest percent of ethnic minority residents (Annelinn, Jaamamoisa, and Ropka tööstusrajoon, with 28%, 39%, and 16.4% of Russians, respectively) were developed mostly during the Soviet era and, for that reason, the predominant typology of buildings were multi-storeyed panel housing blocks. Nowadays, those housing blocks are of poorer quality compared to the single-family house, as described in the literature review, confirming that the majority of ethnic minorities in Tartu, mainly Russians, live in the districts with a declining quality of life.

The World Health Organization (WHO) recommends that there should exist at least 9 m^2 of green space per person, but it recommends 15 m^2 of accessible green space while, the ideal amount, according to the same organisation, would be 50 m^2 [49]. Also, in Brazil, the index of 15 m^2/inhabitant is recommended to promote a good quality of life [81]. The United Nations sustainable development goal 11, regarding sustainable cities, defines, as an indicator, the "adequate open public space in cities", and has as a target, by 2030, to provide universal access to safe, inclusive, and accessible green and public spaces, in particular, for women and children, older persons, and persons with disabilities [82].

Comparing the UN reference point with the values in the Table 4 for Tartu, it is possible to determine that the districts of Ihaste, Jaamamõisa, Karlova, Ropka, Ränilinn, Tammelinn, Vaksali, and Veeriku had a lower area of PGS per inhabitant than that suggested by the World Health Organization. Compared with the same source, only Maarjamõisa, Supilinn, and Tähtvere had over the ideal value of PGS per habitant. However, in this analysis, it was important to note that most districts with low value also had high areas of private gardens, as shown in Table 1.

In Faro, the majority of districts had less than the amount recommended by the World Health Organization [49], such as Figuras district, which had 31.44m² per habitant, below the ideal area of 50 m². According to Barreira et al. [83], some of the reasons for residential satisfaction and city attractiveness were the availability of green areas and opportunities for leisure activities in open-air public space. Green space availability can be related to people's perceived happiness and general health, while having green space nearby appears to reduce the incidence of heart disease, obesity, and depression [84,85]. Moreover, green infrastructure may enhance city resilience to climate change by means of ecosystem services' improvement [20], and assist in climate change adaptation, which is one of the factors for future city growth [86]. According to Panagopoulos [87], city climate change adaptation is a major issue in environmental justice for vulnerable urban populations, like the elderly, minorities, and people of low socioeconomic status, while health benefits and resilience to climate change appear to be stronger for vulnerable groups.

Regarding the accessibility to PGS in Tartu, the results revealed that the districts with less accessible areas of PGS were Jaamamõisa, Variku, Ränilinn, Ihaste, and Veeriku. However, it would be expected that Annelinn and Ropka tööstusrajoon would be also included in the districts with less accessibility of PGS since, according to a survey [88], the inhabitants consider them to be districts with a lower quality living environment, as well as Jaamamõisa and Ränilinn. Still, when critically observing the map in Figure 4, it could be verified that, in Annelinn, which shows a high percentage of accessible area of PGS (52% for the 500 m buffer), a significant part of this area included only few dwellings, while the grid method showed that Annelinn, due to its high population density, had less accessible area of PGS per inhabitant.

While Tartu had a greater area of public green spaces, in total, than Faro, the percentage of accessible area to the PGS was similar in both cities (45%). However, when considering the 300 m distance buffer, Tartu had better access to PGS than Faro (30% instead of 25%). Furthermore, the grid method showed that the accessible area per habitant was higher and more uniform in Tartu than in Faro. The greater accessibility in Tartu could be accounted for by the geographic location and cultural differences. As noted in many studies [5,6], there is a tendency for northern European countries to have greater availability of green spaces, due to a favourable climate compared with Mediterranean countries. Nonetheless, the culture also influences the use of public spaces. For example, Thompson [89] notes that, in Mediterranean countries, there was a long tradition of strolling in the street, while, in Nordic countries, the urban park is the main element of public space, and it plays an important role in social relations. The importance and the use given to PGS can influence the availability of those spaces in the city. However, even if in Mediterranean countries the PGS was less widely used than in the Nordic countries, it was still important to note that the presence and associated ecosystem services of such spaces in the city are essential to securing a high level of quality in urban living.

5.1. Tartu Masterplan (Linna Üldplaneering)

Table 6 summarises the results presented in Tables 4 and 5, and considers the plans for Tartu 2030 and Faro up to 2030. In Tartu, a significant increase in public green space area, around 160 ha, was verified. Consequently, an increase in the accessible area to these spaces is expected. This plan predicts that the total accessible area in Tartu will increase from 30% to 68%, and from 45% to 86%, using a 300 m and 500 m distance buffer, respectively.

Table 6. Comparison between the current accessibility to the public green spaces (PGS) and after implementing Tartu Masterplan and Faro new green infrastructure proposal.

City	Area (ha)	PGS (ha)	PGS m²/inhabitant	Accessible Area 300 m		Accessible Area 500 m		Non-Accessible Area 300 m		Non-Accessible Area 500 m	
Tartu	3897	202	20.8	1153 ha	30%	1747 ha	45%	2744 ha	70%	2150 ha	55%
Tartu Masterplan	3897	355	36.5	2635 ha	68%	3346 ha	86%	1262 ha	32%	551 ha	14%
Faro	711	26	6.2	183 ha	25%	321 ha	45%	528 ha	75%	390 ha	55%
Faro proposal	711	92	22.3	522 ha	73%	628 ha	88%	189 ha	17%	83 ha	12%

On the implementation of the Tartu Masterplan by 2030, the districts with a high proportion of accessible area to PGS will be Supilinn, Jaamamõisa, Kesklinn, and Ränilinn, while the districts with a lower proportion will be Tammelinn, Veeriku, and Ropka tööstusrajoon, for both distance buffers.

Nevertheless, the data per district shows that the increase in the PGS and, consequently, in the percentage of accessible area, would result primarily from two transformations in the landscape of the city: (1) an empty area, without any existing function, could be transformed into a PGS of high quality, or (2) an existing PGS of low quality could be transformed into a new one of high quality.

Variku and Ihaste are examples of those planned transformations (Figure 6). In Ihaste, the area of PGS should increase by 46 ha; however, this area represents existing green spaces of low quality and without mobility conditions. In turn, the PGS area of Variku should increase by 10 ha, which would be occupied by a new park in an area that currently does not have any function. According to Tartu Masterplan, this new public space will include a dog walking area and recreational, sports, and cultural facilities.

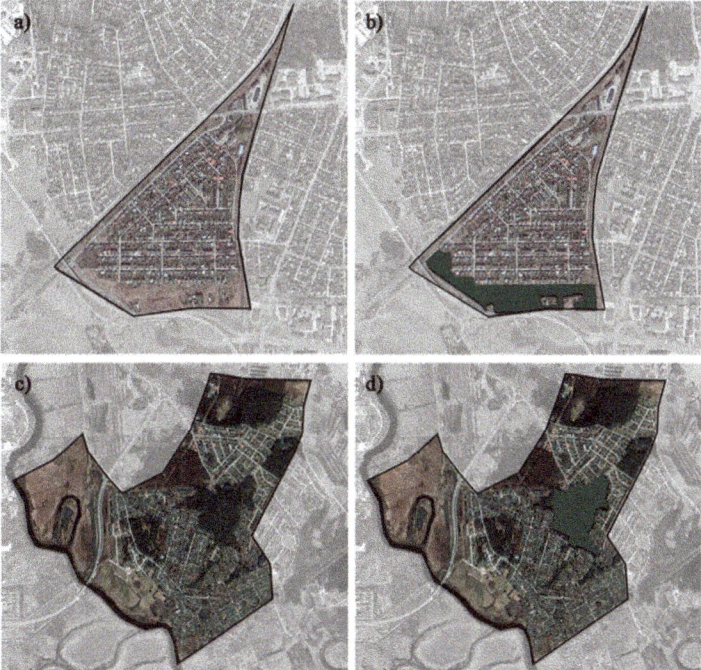

Figure 6. Proposal for new green infrastructure using the "Tartu Linna Üldplaneering", a general plan for 2030. (**a**) Variku before; (**b**) Variku after; (**c**) Ihaste before; (**d**) Ihaste after. (Source: Satellite image from Google Earth).

5.2. Proposal for Faro

The results of the present study revealed that some districts in Faro were without accessibility to PGS, which allows us to confirm that there was an unequal distribution of PGS in this city. This fact justifies the necessity of suggesting some strategies for increasing the amount of PGS and, consequently, promoting the equal distribution of these areas across all Faro's districts. The following map (Figure 7) identifies the location of the main problem areas in the city, and proposes solutions for those issues.

The most important intervention is the development of an urban park in Penha and the redevelopment of the industrial area, to promote the connection between inhabitants and the Ria Formosa. For the most compact districts—1, 2, and 3—without space to develop new green areas, construction of green roofs and green walls on public buildings, together with the planting of trees along streets, was proposed to promote the connection of existing green areas and the continuity of ecological flows. For disaster risk reduction and climate change adaptation, soil restoration and substitution of the pavement with permeable materials, in stream margins with risk of flash flooding, was proposed.

In the Old Town area, included in district 1, the rehabilitation of private abandoned yards and urban agriculture allotments was proposed, to establish a connection between these areas with the existing PGS enhancing the city resilience [90]. Those abandoned private yards, being located in a historic district and close to the "Ria Formosa", have a high ecological and recreational potential, especially for the inhabitants of the old city that are mostly elderly. It was also proposed to convert the current railway line to a bike path and green corridor. These proposals must integrate the ideas, expectations, and suggestions of the inhabitants [91] and of the visitors [92], which will be part of future research.

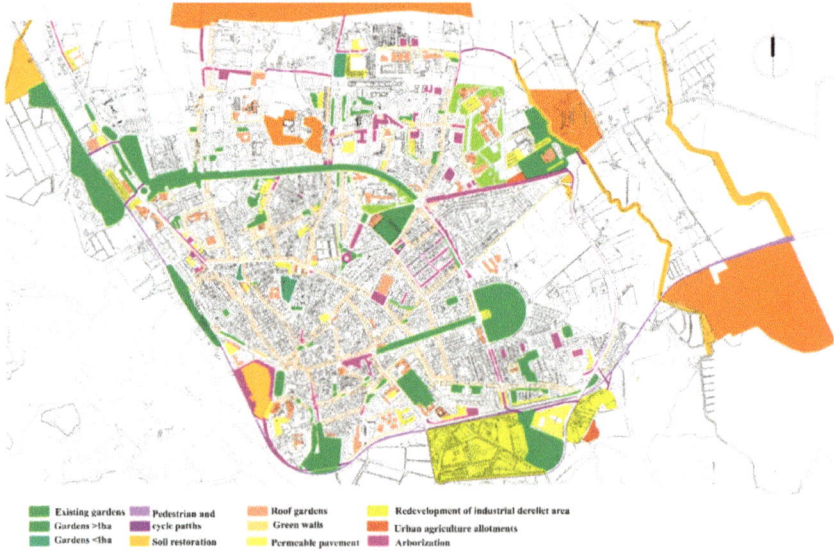

Figure 7. Proposals for new green infrastructure in Faro.

5.3. Green Gentrification

Taking the definition of green gentrification, as discussed earlier in the paper, into account, and comparing the Tartu and Faro cases, in our view, this phenomenon is more likely to occur in Faro than in Tartu, since the need to develop new green spaces is greater in Faro than in Tartu. In addition, Tartu presents, for the most part, a typology of buildings that allows for the presence of private gardens as an alternative to the need for PGS by the residents of these single-family houses.

In Faro, the abandoned industrial area, included in district 5, has been set aside for an urban redevelopment project, as it has direct contact with the Ria Formosa landscape, is close to the city centre and the airport, and forms an extensive area. Municipal plans refer to the implementation of a future project to include a marina and a tourist-residential area with extensive PGS, that will occupy 29 ha. The municipal authority believes that this project will not compromise the operation of the current commercial port and the high environmental qualities that characterise that landscape, meanwhile estimating that this project has a total potential value of 300 million euros for the city [93]. Meanwhile, this highly polluted area, with residents mostly of Roma ethnicity living in state housing, represents a case of social exclusion in the city, with problems of poverty, crime, and lack of security. It is expected that the implementation of a large urban regeneration project in this residential district, inserted in this old industrial area since 1970, will directly affect the life of the residents of the state housing, since they would be forced to be relocated to another part of the city. Thus, this area has a high likelihood of green gentrification.

In the case of the Tartu, it is thought that the neighbourhoods built during the Soviet period are where the phenomenon of eco-gentrification may exist since, as stated in the literature, they present poorer conditions than others [94]. However, the fact that there are a wide variety of areas suitable for the construction of new public green spaces, and because Tartu is considered a fairly balanced city, the likelihood of green gentrification occurring is reduced.

A limitation of this study is the fact that gentrification might be a result, as well as a cause, of green space development. As our data do not allow the assessment of causal relationships over time between gentrification and new green space development, it remains difficult to elaborate further on the situation. It was not possible to evaluate the existence of green gentrification, in part, due to lack of data about the income of households, over time, per district, for both cities. Also, in Faro, although an area with risk of gentrification was identified, it was not possible to develop a method of calculating this phenomenon. For future research, to evaluate the green gentrification phenomenon, we suggested that additional socioeconomic data should be obtained and applied in a Markov Model of Urban Change in time and space, as was done by Royall and Wortmann [95].

6. Conclusions

Urban PGS perform important functions that directly affect the quality of life and wellbeing of urban inhabitants. They are considered key elements in improving the quality of life and creating an appropriate framework for sustainable cities [96]. For this reason, such public spaces should be equitably distributed throughout the city, in order to provide ecosystem services to the entire population. Accessibility and availability of PGS can be valuable indicators of environmental injustice in neighbourhoods within a city, information which may help to promote future urban regeneration projects in areas with the highest needs. The quantitative indicators used to estimate availability and assess urban green infrastructure can be applied widely in comparisons of any city, or of districts within city [97]. Meanwhile, one limitation of the study was the inability to consider the quality of green space, as a result of lack of data.

Based on the analysis of urban growth and indicators related to the distribution of PGS, it was found that, in both cities, there was a relationship to the ethnicity of the inhabitants. Moreover, the study between Tartu and Faro showed that the geographical location within Europe can influence the urban space availability with 6.2 m^2/inhabitant in Faro and 20.8 m^2/inhabitant in Tartu. Meanwhile, the accessibility to PGS was 45% of residences in both cities, because Tartu is low density and Faro is a compact city.

Tartu ethnic minorities, mainly Russians that live in areas developed during the Soviet period (Jaamamõisa, Ropka tööstusrajoon, and Annelinn), were those who live in districts classified as having lower quality of the living environment [88]. Meanwhile, the walkability distance method revealed that districts Variku, Ränilinn, and Veeriku also had low accessibility while, for Annelinn, all indicators were close to the city average (14 m^2/inhabitant, 32% accessibility within 300 m, and 52% within

500 m). From the above, it is concluded that there is a strong relationship between the ethnicity, type of urbanisation, and the quality of the living environment, that includes the availability and accessibility to PGS, but is not confirmed in every district. Despite this, the degree of environmental injustice in Tartu was lower than Faro.

The minorities of Faro, in particular, the Roma communities, were located in districts without (Penha 1.22 m^2/inhabitant) or with low access to public green spaces (Bom João 4.76 m^2/inhabitant). Furthermore, those neighbourhoods may suffer gentrification processes in the future, due to the implementation of large-scale urban regeneration projects aiming to satisfy the tourist demands of the region. Meanwhile, it was not possible to evaluate the green gentrification phenomenon, because this is a long process, beyond the period of this research, and it was not clear if gentrification might be a result, as well as a cause, of green space development.

In Faro, Figuras, which is the most recently developed district and with inhabitants of higher socioeconomic status, showed greater availability and accessibility to green areas (31.44 m^2/inhabitant, 35% accessibility within 300 m, and 51% within 500 m). This district includes both of the new PGS established in Faro in recent decades (Parque Ribeirinho and the Parque de Lazer). In Faro, being one of the Portuguese cities with the least green space per inhabitant, it is an urgent task to focus on planning new green infrastructure, integrating the needs and opinions of residents into proposals, in order to increase the accessibility to PGS and secure environmental justice in the city.

The comparison between Tartu and Faro suggests that the geographical location influenced the availability and quality of green areas in the city, as well as social cohesion itself. In addition, it is expected that the implementation of "Tartu Linna Üldplaneering", which reflects a scenario for the city's public green spaces by 2030, will considerably improve the city's environmental justice. The proposal for new green infrastructure in Faro may decrease the verified environmental injustice.

Accessibility to PGS in cities, and the identification of the most problematic areas, should be integrated into sustainable urban planning proposals. The outcome of this study could provide good advice for balancing green space distribution within city neighbourhoods in similar cities of other countries. In addition, future actions should be conducted with the aim of monitoring the long-term ecosystem services provided by green spaces, and for early identification of green gentrification risks.

Author Contributions: Conceptualization, C.d.S.S. and T.P.; Methodology, C.d.S.S.; Validation, C.d.S.S., T.P. and S.B.; Formal Analysis, C.d.S.S.; Investigation, C.d.S.S. and I.V.; Writing—Original Draft Preparation, C.d.S.S. and T.P.; Writing—Review & Editing, T.P. and S.B.; Supervision, T.P. and S.B.

Funding: This research was funded by the Foundation for Science and Technology grant number PTDC/GES-URB/31928/2017 and co-funded by the Erasmus+ programme of the European Union.

Acknowledgments: This paper was financed by the FCT-Foundation for Science and Technology through project PTDC/GES-URB/31928/2017 "Improving life in a changing urban environment through Biophilic Design". We also thank Gloria Niin, Janar Racet and Sille Tiitsmaa for providing data.

Conflicts of Interest: The authors declare no conflict of interest. The founding sponsors had no role in the design of the study; in the collection, analyses, or interpretation of data; in the writing of the manuscript, and in the decision to publish the results.

References

1. Artmann, M.; Kohler, M.; Meinel, G.; Gan, J.; Ioja, I.C. How smart growth and green infrastructure can mutually support each other—A conceptual framework for compact and green cities. *Ecol. Indic.* **2017**. [CrossRef]

2. Hansen, R.; Olafsson, A.S.; van der Jagt, A.P.N.; Rall, E.; Pauleit, S. Planning multifunctional green infrastructure for compact cities: What is the state of practice? *Ecol. Indic.* **2017**. [CrossRef]

3. Zwierzchowska, I.; Hof, A.; Ioja, I.C.; Mueller, C.; Poniży, L.; Breuste, J.; Mizgajski, A. Multi-scale assessment of cultural ecosystem services of parks in Central European cities. *Urban For. Urban Green.* **2018**, *30*, 84–97. [CrossRef]

4. Iojă, I.C.; Osaci-Costache, G.; Breuste, J.; Hossu, C.A.; Grădinaru, S.R.; Onose, D.A.; Niță, M.R.; Skokanová, H. Integrating urban blue and green areas based on historical evidence. *Urban For. Urban Green.* **2018**, *3*, 217–225. [CrossRef]

5. Fuller, R.A.; Gaston, K.J. The scaling of green space coverage in European cities. *Boil. Lett.* **2009**, *5*. [CrossRef] [PubMed]

6. Kabisch, N.; Strohbach, M.; Haase, D.; Kronenberg, J. Urban green space availability in European cities. *Ecol. Indic.* **2016**, *70*, 586–596. [CrossRef]

7. Park, Y.M.; Kwan, M.P. Multi-contextual segregation and environmental justice research: Toward fine-scale spatiotemporal approaches. *Int. J. Environ. Res. Public Health* **2017**, *14*, 1205. [CrossRef] [PubMed]

8. Hoffimann, E.; Barros, H.; Ribeiro, A.I. Socioeconomic inequalities in green space quality and accessibility-evidence from a Southern European City. *Int. J. Environ. Res. Public Health* **2017**, *14*, 916. [CrossRef] [PubMed]

9. Herman, K.; Sbarcea, M.; Panagopoulos, T. Creating green space sustainability through low-budget and upcycling strategies. *Sustainability* **2018**, *10*, 1857. [CrossRef]

10. Terkenli, S.T.; Zivojinovic, I.; Tomićević-Dubljević, J.; Panagopoulos, T.; Straupe, I.; Toskovic, O.; Kristianova, K.; Straigyte, L.; O'Brien, L.; Bell, S. Recreational Use of Urban Green Infrastructure: The Tourist's Perspective. In *The Urban Forest*; Pearlmutter, D., Calfapietra, C., Samson, R., O'Brien, L., Krajter Ostoić, S., Sanesi, G., Alonso del Amo, R., Eds.; Springer: Cham, Switzerland, 2017; pp. 191–216.

11. Madureira, H.; Nunes, F.; Oliveira, J.V.; Cormier, L.; Madureira, T. Urban residents' beliefs concerning green space benefits in four cities in France and Portugal. *Urban For. Urban Green.* **2015**, *14*, 56–64. [CrossRef]

12. Revez, I. A Cidade de Faro Está 'Doente', Asfixiada Pela Falta de Áreas Verdes Para a População. Available online: https://www.publico.pt/2011/11/21/jornal/a-cidade-de-faro-esta-doente-asfixiada-pela-falta-de-areas-verdes-para-a-populacao-23462408 (accessed on 24 January 2018).

13. Benedict, M.A.; McMahon, E.T. Green infrastructure: Smart conservation for the 21st century. *Renew. Resour. J.* **2002**, *20*, 12–17.

14. Weber, T.; Wolf, J. Maryland's green infrastructure—Using landscape assessment tools to identify a regional conservation strategy. *Environ. Monit. Assess.* **2000**, *63*, 265–277. [CrossRef]

15. Ahern, J. Green infrastructure for cities: The spatial dimension. In *Cities of the Future: Towards Integrated Sustainable Water and Landscape Management*; IWA Publishing: London, UK, 2007.

16. Little, C.E. *Greenways for America*; JHU Press: Baltimore, MD, USA, 2007.

17. European Commission. *Building a Green Infrastructure for Europe*; European Commission: Luxemburg, 2013.

18. Berte, E.; Panagopoulos, T. Enhancing city resilience to climate change by means of ecosystem services improvement: A SWOT analysis for the city of Faro, Portugal. *Int. J. Urban Sustain. Dev.* **2014**, *6*, 241–253. [CrossRef]

19. Millennium Ecosystem Assessment. *Ecosystems and Human Well-Being: Synthesis*; Island Press: Washington, DC, USA, 2005.

20. Maas, J.; Verheij, R.A.; Groenewegen, P.P.; Vries, S.; Spreeuwenberg, P. Green space, urbanity, and health: How strong is the relation? *J. Epidemiol. Community Health* **2006**, *60*, 587–592. [CrossRef] [PubMed]

21. Panagopoulos, T. Linking forestry, sustainability and aesthetics. *Ecol. Econ.* **2009**, *68*, 2485–2489. [CrossRef]

22. Kabisch, N.; Haase, D. Green spaces of European cities revisited for 1990–2006. *Landsc. Urban Plan.* **2013**, *110*, 113–122. [CrossRef]

23. Costanza, R.; d'Arge, R.; de Groot, R.; Farber, S.; Grasso, M.; Hannon, B.; Limburg, K.; Naeem, S.; ONeil, R.V.; Paruelo, J.; et al. The value of the world's ecosystem services and natural capital. *Nature* **1997**, *387*, 253–260. [CrossRef]

24. Kabisch, N.; Haase, D. Green justice or just green? Provision of urban green spaces in Berlin, Germany. *Landsc. Urban Plan.* **2014**, *122*, 129–139. [CrossRef]

25. Quastel, N. Political ecologies of gentrification. *Urban Geogr.* **2009**, *30*, 694–725. [CrossRef]

26. Curran, W.; Hamilton, T. Just green enough: Contesting environmental gentrification in Greenpoint, Brooklyn. *Local Environ.* **2012**, *17*, 1027–1042. [CrossRef]

27. Byrne, J.; Wolch, J.; Zhang, J. Planning for environmental justice in an urban national park. *J. Environ. Plan. Manag.* **2009**, *52*, 365–392. [CrossRef]

28. Gould, K.A.; Lewis, T.L. *Green Gentrification. Urban Sustainability and the Struggle for Environmental Justice*; Routledge: New York, NY, USA, 2017.

29. Taylor, W.C.; Poston, C.; Jones, L.; Kraft, M.K. Environmental justice: Obesity, physical activity, and healthy eating. *J. Phys. Act. Health* **2006**, 3, 30–54. [CrossRef] [PubMed]
30. Downey, L.; Hawkins, B. Race, income, and environmental inequality in the United States. *Sociol. Perspect.* **2008**, 51, 759–781. [CrossRef] [PubMed]
31. Rigolon, A. Parks and young people: An environmental justice study of park proximity, acreage, and quality in Denver, Colorado. *Landsc. Urban Plan.* **2017**, 165, 73–83. [CrossRef]
32. Laurent, E. Environmental Justice and Environmental Inequalities: A European Perspective. Available online: http://inis.iaea.org/Search/search.aspx?orig_q=RN:47006484 (accessed on 12 May 2018).
33. Schlosberg, D. Reconceiving environmental justice: Global movements and political theories. *Environ. Politics* **2004**, 13, 517–540. [CrossRef]
34. Shen, Y.; Sun, F.; Che, Y. Public green spaces and human wellbeing: Mapping the spatial inequity and mismatching status of public green space in the Central City of Shanghai. *Urban For. Urban Green.* **2017**, 27, 59–68. [CrossRef]
35. Wolch, J.; Byrneb, J.; Newell, J.P. Urban green space, public health, and environmental justice: The challenge of making cities 'just green enough'. *Landsc. Urban Plan.* **2014**, 125, 234–244. [CrossRef]
36. Wüstemann, H.; Kalisch, D.; Kolbe, J. Access to urban green space and environmental inequalities in Germany. *Landsc. Urban Plan.* **2017**, 164, 124–131. [CrossRef]
37. Raudsepp, M.; Heidmets, M.; Kruusvall, J. Environmental Justice and Sustainability in Post-Soviet Estonia. In *Environmental Justice and Sustainability in the Former Soviet Union*; Agyeman, J., Ogneva-Himmelberger, Y., Eds.; The MIT Press: London, UK, 2009.
38. Steger, T. *Making the Case of Environmental Justice in Central & Eastern Europe*; Center for Environmental Policy and Law: Budapest, Hungary, 2007.
39. Boyce, J.K.; Zwickl, K.; Ash, M. Measuring environmental inequality. *Ecol. Econ.* **2016**, 124, 114–123. [CrossRef]
40. Fan, P.; Xu, L.; Yue, W.; Chen, J. Accessibility of public urban green space in an urban periphery: The case of Shanghai. *Landsc. Urban Plan.* **2017**, 165, 177–192. [CrossRef]
41. Lindholst, A.C.; Konijnendijk, C.C.; Kjøller, C.P.; Sullivan, S.; Kristoffersson, A.; Fors, H.; Nilsson, K. Urban green space qualities reframed toward a public value management paradigm: The case of the Nordic Green Space Award. *Urban For. Urban Green.* **2016**, 17, 166–176. [CrossRef]
42. Rojas, C.; Páez, A.; Barbosa, O.; Carrasco, J. Accessibility to urban green spaces in Chilean cities using adaptive thresholds. *J. Transp. Geogr.* **2016**, 57, 227–240. [CrossRef]
43. Chiesura, A. The role of urban parks for the sustainable city. *Landsc. Urban Plan.* **2004**, 68, 129–138. [CrossRef]
44. Coles, R.W.; Bussey, S.C. Urban forest landscapes in the UK—Progressing the social agenda. *Landsc. Urban Plan.* **2000**, 52, 181–188. [CrossRef]
45. Van Herzele, A.; Wiedemann, T. A monitoring tool for the provision of accessible and attractive urban green spaces. *Landsc. Urban Plan.* **2003**, 63, 109–126. [CrossRef]
46. Comber, A.; Brunsdon, C.; Green, E. Using a GIS-based network analysis to determine urban greenspace accessibility for different ethnic and religious groups. *Landsc. Urban Plan.* **2008**, 86, 103–114. [CrossRef]
47. Handley, J.; Pauleit, S.; Slinn, P.; Barber, A.; Baker, M.; Jones, C. *Accessible Natural Green Space Standards in Towns and Cities: A Review and Toolkit for Their Implementation*; English Nature: Peterborough, UK, 2003.
48. Magalhães, M.R. *Espaços Verdes Urbanos*; Direcção-Geral do Ordenamento do Território: Lisbon, Portugal, 1992.
49. World Health Organization. Urban Planning, Environment and Health: From Evidence to Policy Action. Available online: http://www.euro.who.int/__data/assets/pdf_file/0004/114448/E93987.pdf?ua=1 (accessed on 22 April 2018).
50. Maroko, A.R.; Maantay, J.A.; Sohler, N.L.; Grady, K.L.; Arno, P.S. The complexities of measuring access to parks and physical activity sites in New York City: A quantitative and qualitative approach. *Int. J. Health Geogr.* **2009**, 8, 34. [CrossRef] [PubMed]
51. Talen, E.; Anselin, L. Assessing spatial equity: An evaluation of measures of accessibility to public playgrounds. *Environ. Plan. A* **1998**, 30, 595–613. [CrossRef]
52. Timperio, A.; Ball, K.; Salmon, J.; Roberts, R.; Crawford, D. Is availability of public open space equitable across areas? *Health Place* **2007**, 13, 335–340. [CrossRef] [PubMed]
53. Moore, L.V.; Diez Roux, A.V.; Evenson, K.R.; McGinn, A.P.; Brines, S.J. Availability of recreational resources in minority and low socioeconomic status areas. *Am. J. Prev. Med.* **2008**, 34, 16–22. [CrossRef] [PubMed]

54. Dai, D. Racial/ethnic and socioeconomic disparities in urban green space accessibility: Where to intervene? *Landsc. Urban Plan.* **2011**, *102*, 234–244. [CrossRef]

55. Smith, N. Gentrification and uneven development. *Econ. Geogr.* **1982**, *58*, 139–155. [CrossRef]

56. Smith, N. New globalism, new urbanism: Gentrification as global urban strategy. *Antipode* **2002**, *34*, 427–450. [CrossRef]

57. Gould, K.; Lewis, T. The Environmental Injustice of Green Gentrification: The Case of Brooklyn's Prospect Park. In *The World in Brooklyn: Gentrification, Immigration, and Ethnic Politics in a Global City*; DeSena, J., Shortell, T., Eds.; Lexington Books: Lanham, MD, USA, 2012; pp. 113–146.

58. Dooling, S. Ecological gentrification: A research agenda exploring justice in the city. *Int. J. Urban Reg. Res.* **2009**, *33*, 621–639. [CrossRef]

59. Bentley, R.; Baker, E.; Mason, K. Cumulative exposure to poor housing affordability and its association with mental health in men and women. *J. Epidemiol. Community Health* **2012**, *66*, 761–766. [CrossRef] [PubMed]

60. Checker, M. Wiped out by the 'greenwave': Environmental gentrification and the paradoxical politics of urban sustainability. *City Soc.* **2011**, *23*, 210–229. [CrossRef]

61. O'Brien, L.; De Vreese, R.; Atmis, E.; Olafsson, A.S.; Sievänen, T.; Brennan, M.; Sánchez, S.; Panagopoulos, T.; de Vries, S.; Kern, M.; et al. Social and Environmental Justice: Diversity in Access to and Benefits from Urban Green Infrastructure—Examples from Europe. In *The Urban Forest*; Pearlmutter, D., Calfapietra, C., Samson, R., O'Brien, L., Krajter Ostoić, S., Sanesi, G., Alonso del Amo, R., Eds.; Springer: Cham, Switzerland, 2017; pp. 153–190.

62. Walker, S. Urban agriculture and the sustainability fix in Vancouver and Detroit. *Urban Geogr.* **2016**, *37*, 163–182. [CrossRef]

63. Colasanti, K.J.A.; Hamm, M.W.; Litjens, C.M. The city as an 'agricultural powerhouse'? Perspectives on expanding urban agriculture from Detroit, Michigan. *Urban Geogr.* **2012**, *33*, 348–369. [CrossRef]

64. Kloek, M.E.; Buijs, A.E.; Boersema, J.J.; Schouten, M.G.C. 'Nature lovers', 'Social animals', 'Quiet seekers' and 'Activity lovers': Participation of young adult immigrants and non-immigrants in outdoor recreation in the Netherlands. *J. Outdoor Recreat. Tour.* **2015**, *12*, 47–58. [CrossRef]

65. Demko, G.J.; Regulska, J. Socialism and Its Impact on Urban Processes and the City. *Urban Geogr.* **1987**, *8*, 289–292. [CrossRef]

66. Smith, D.M. The Socialist City. In *Cities after Socialism*; Andrusz, G., Harloe, M., Szelenyi, I., Eds.; Wiley-Blackwell: Oxford, UK, 1996.

67. Andrusz, G.; Harloe, M.; Szelenyi, I. *Cities After Socialism: Urban and Regional Change and Conflict in Post-Socialist Societies*; Blackwell: Oxford, UK, 1996.

68. Vetik, R.; Helemäe, J. *The Russian Second Generation in Tallinn and Kohtla-Järve: The TIES Study in Estonia*; Amsterdam University Press: Amsterdam, The Netherlands, 2011.

69. Kulu, H. Housing differences in the late Soviet city: The case of Tartu, Estonia. *Int. J. Urban Reg. Res.* **2003**, *27*, 897–911. [CrossRef]

70. Kährik, A.; Tammaru, T. Soviet prefabricated panel housing estates: Areas of continued social mix or decline? The case of Tallinn. *Hous. Stud.* **2010**, *25*, 201–219. [CrossRef]

71. Gehl, J. *Cities for People*; Island Press: Copenhagen, Denmark, 2010.

72. Congress for the New Urbanism. What is New Urbanism? Available online: https://www.cnu.org/resources/what-new-urbanism (accessed on 15 May 2017).

73. Steiner, F.R. Landscape ecological urbanism: Origins and trajectories. *Landsc. Urban Plan.* **2011**, *100*, 333–337. [CrossRef]

74. Thompson, I. Ten tenets and six questions for landscape urbanism. *Landsc. Res.* **2012**, *37*, 7–26. [CrossRef]

75. Sousa Silva, C.; Lackóová, L.; Panagopoulos, T. Applying sustainability techniques in eco-industrial parks. *WIT Trans. Ecol. Environ.* **2017**, *210*, 135–145. [CrossRef]

76. Karanikola, P.; Panagopoulos, T.; Tampakis, S.; Tsantopoulos, G. Cycling as a smart and green mode of transport in small touristic cities. *Sustainability* **2018**, *10*, 268. [CrossRef]

77. Smart Growth Network. Available online: http://smartgrowth.org/what-is-smart-growth/ (accessed on 8 June 2018).

78. Hess, D.B.; Tammaru, T.; Leetmaa, K. Ethnic differences in housing in post-Soviet Tartu, Estonia. *Cities* **2012**, *29*, 327–333. [CrossRef]

79. Santos, S.R.; Marques, J.F. O Rendimento Social de Inserção e os beneficiários ciganos: O caso do concelho de Faro. *Sociologia* **2014**, *4*, 37–56.
80. Santos, S.A. Tenho a Noite e o Dia e Não Tenho Nada: O Rendimento Social de Inserção e os Beneficiários Ciganos: O Caso Do Concelho de Faro. Available online: https://sapientia.ualg.pt/handle/10400.1/35312013 (accessed on 13 May 2018).
81. Harder, I.C.F.; Ribeiro, R.C.S.; Tavares, A.R. Green area and vegetation cover indexes for commons in the city of Vinhedo, SP. *Arvore* **2006**, *30*, 277–282. [CrossRef]
82. United Nations. Available online: http://www.undp.org/content/undp/en/home/sustainable-development-goals/goal-11-sustainable-cities-and-communities/targets.html (accessed on 13 March 2018).
83. Barreira, AP.; Nunes, L.C.; Guimaraes, M.H.; Panagopoulos, T. Satisfied but thinking about leaving: The reasons behind residential satisfaction and residential attractiveness in shrinking Portuguese cities. *Int. J. Urban Sci.* **2018**. [CrossRef]
84. Saw, L.E.; Lim, F.K.S.; Carrasco, L.R. The relationship between natural park usage and happiness does not hold in a tropical city-state. *PLoS ONE* **2015**, *10*, e0133781. [CrossRef] [PubMed]
85. Hillsdon, M.; Panter, J.; Foster, C.; Jones, A. The Relationship between access and quality of urban green space with population physical activity. *Public Health* **2006**, *120*, 1127–1132. [CrossRef] [PubMed]
86. Barreira, A.P.; Ramalho, J.J.S.; Panagopoulos, T.; Guimarães, M.E. Factors driving the population growth and decline of Portuguese cities. *Growth Chang.* **2017**, *48*, 853–868. [CrossRef]
87. Panagopoulos, T. Urban adaptation to climate change: The role of the Landscape Architecture. In *Natural and Man-Made Hazard Impact on Urban Areas*; Dan, M.B., Anghelache, M.A., Eds.; Editura Universitară Ion Mincu: Bucurest, Romania, 2017; pp. 13–17.
88. Küsitlusuuring Tartu Ja Tartlased 2013. Available online: https://www.tartu.ee/et/uurimused/kusitlusuuring-tartu-ja-tartlased-2013 (accessed on 17 January 2018).
89. Thompson, C.W. Urban open space in the 21st century. *Landsc. Urban Plan.* **2002**, *60*, 59–72. [CrossRef]
90. Panagopoulos, T.; Jankovska, I.; Dan, M.B. Urban Green Infrastructure: The Role of Urban Agriculture in City Resilience. *Urban. Arhit. Constr.* **2018**, *9*, 55–70.
91. Karanikola, P.; Panagopoulos, T.; Tampakis, S.; Karipidou-Kanari, A. A perceptual study of users' expectations of urban green infrastructure in Kalamaria, municipality of Greece. *Manag. Environ. Qual. Int. J.* **2016**, *27*, 568–584. [CrossRef]
92. Karanikola, P.; Panagopoulos, T.; Tampakis, S. Weekend visitors' views and perceptions at an urban national forest park of Cyprus during summertime. *J. Outdoor Recreat. Tour.* **2017**, *17*, 112–121. [CrossRef]
93. Claro, R. Marina Na Zona da Horta da Areia Pode Garantir Requalificação. Available online: http://www.postal.pt/2016/04/marina-na-zona-da-horta-da-areia-pode-garantir-requalificacao/ (accessed on 8 December 2017).
94. Kährik, A. Housing privatisation in the transformation of the housing system—The case of Tartu, Estonia. *Nor. J. Geogr.* **2000**, *54*, 2–11. [CrossRef]
95. Royall, E.; Wortmann, T. Finding the State Space of Urban Regeneration: Modeling Gentrification as a Probabilistic Process using k-Means Clustering and Markov Models. In Proceedings of the 2015 14th International Conference on Computers in Urban Planning and Urban Management (CUPUM), Cambridge, MA, USA, 7–10 July 2015; p. 275.
96. Panagopoulos, T.; González Duque, J.A.; Bostenaru Dan, M. Urban planning with respect to environmental quality and human well-being. *Environ. Pollut.* **2016**, *208*, 137–144. [CrossRef] [PubMed]
97. Badiu, D.L.; Ioja, C.I.; Patroescu, M.; Breuste, J.; Artmann, M.; Nita, M.; Gradinaru, S.; Hossu, C.; Onose, D.A. Is urban green space per capita a valuable target to achieve cities' sustainability goals? Romania as a case study. *Ecol. Indic.* **2016**, *70*, 53–66. [CrossRef]

Article

Residents' Perception of Informal Green Space—A Case Study of Ichikawa City, Japan

Minseo Kim [1,*], Christoph D. D. Rupprecht [2] and Katsunori Furuya [1]

[1] Department of Environment Science and Landscape Architecture, Graduate School of Horticulture, Chiba University, Chiba 271-8510, Japan; k.furuya@faculty.chiba-u.jp

[2] FEAST Project, Research Institute for Humanity and Nature, Kyoto 6038047, Japan; crupprecht@chikyu.ac.jp

* Correspondence: k.minseo@chiba-u.jp; Tel.: +81-70-2679-8443

Received: 10 August 2018; Accepted: 31 August 2018; Published: 4 September 2018

Abstract: Urban green space (UGS) has been proven to be essential for improving the health of residents. Local governments thus need to provide attractive UGS to enhance residents' wellbeing. However, cities face spatial and finanical limitations in creating and managing UGS. As a result, greening plans often fail or are postponed indefinitely. To evaluate whether informal urban green space (IGS) can supplement existing UGS, we conducted a questionnaire survey of 567 residents in Ichikawa (Japan), a city currently providing only 3.43 m^2 green space per capita. In particular, we analyzed how residents' existing green space activities affect IGS perception, as it may be difficult to recognize IGS as greenery because it is not an officially recognized space for recreation. Results show that residents took a favorable stance towards IGS, but perception differs depending on their green environment exposure. Residents who are frequently exposed to green environments in their daily lives highly recognized the environmental improvement aspects of IGS and significantly perceived spatial accessibility as an advantage of IGS. Willingness to participate in conservation activities of UGS was linked with a likelihood of recognizing IGS as UGS. Our results encourage understanding IGS as supplementary green space taking into account the attitude of residents to UGS, and contribute to introducing the IGS discourse into green space planning.

Keywords: vacant land; street verges; spontaneous vegetation; postal questionnaire; Asia; Japan; recreation

---

1. Introduction

Urbanization throughout the world has led an increasing proportion of the population living in cities. The United Nations expects that 68% of the world's population will live in urban areas by 2050 [1]. As urbanization progresses and the urban proportion of the population increases, residents living in areas with paved environments often experience limited nature contact and increased exposure to noise and air pollution [2]. Therefore, many studies have focused their attention on urban green spaces (UGS), such as urban parks, forests, gardens, etc., to improve urban dwellers' quality of life and the urban environment. UGS plays a role in providing nature contact directly or indirectly in urban areas, supporting people's physical health and well-being. This support positively affects human mental health, including stress reduction [3–5]. In addition, UGS can also enhance social cohesion and attachment to a place, as well as encourage outdoor activities [6,7]. Therefore, the perception that UGS is an essential element in determining the quality of life of residents is well established. Local or national governments have thus created UGS as part of urban planning strategies to improve or support urban residents' wellbeing and urban environment [8–10]. However, building and managing new parks in the urban area places a financial burden on budgets [11]. This cost associated with public projects, such as creating an urban park, is particularly noticeable in countries, like Japan, where economic growth has reached its peak and cities have begun to shrink [12,13]. The national budget of

Japan for promoting public infrastructures, which includes the creation and maintenance of UGS, has been steadily declining since it peaked in 1997 [14].

Urban or green space planning mostly focuses on the formal and generally acknowledged UGS, including parks, forests, public gardens, and cemeteries. These UGS are highly managed using officially collected data, which provide the basis for extensive research [15]. However, urban spaces go through cycles of planning and (re)development repeatedly and regularly, which can generate spatial by-products, such as vacant lands, wastelands, brownfields, and arable, which could be recognized as leftover spaces [16]. These are generated not as a result of degradation and destruction, but as a result of differences in time as spatial byproducts of policy action [17]. Such spaces range from vacant lots in marginal areas to tiny cracks in between paved lanes. Previous studies have challenged the orthodox ideas of planing through discourses, such as 'place-making' in the contemporary city, in the context of these informal spaces [16,18]. Physically, these spaces are mainly covered with spontaneous vegetation of native or exotic species, mixed with construction rubble or subsoil, with little maintenance [19].

Recent research has drawn attention to reconsidering the possibility of formalizing these spaces to contribute to urban sustainability as green infrastructure [20–24], and provides evidence that these spaces can be valuable as green space and can meet the conditions necessary for recreational use [12,25–27]. Rupprecht and Byrne [25,28] call these spaces informal urban green space (IGS) and define IGS as a space with a history of strong artificial disturbance and spontaneous vegetation occupying some or all of the space. They classified IGS into nine types: Street verges, lots, gap, railway, brownfields, waterside, structural, microsite, and powerline. Furthermore, Rupprecht [12] proposed a participatory IGS management approach based on a survey of residents' perceptions in four representative shrinking cities in Japan. IGS is valued by residents similar to UGS, particularly in regard to the opportunity to access nature in urban areas [26]. However, a recent review found that the biodiversity literature is critically biased in its focus on urban forests or parks and its neglect of IGS [27]. Despite studies' efforts to enrich the discourse about green spaces, like IGS, that are not included in the formal classification and to work towards empirical management systems, it is still not recognized by stakeholders in urban planning. In the evolving discourse on IGS, of course, proposed solutions that distinguish green spaces in binaries, such as informal and formal, and focus solely on scientific-ecological arguments may not sufficiently capture the dynamics between humans and nature in urban areas [28,29]. Further research is thus needed on how residents perceive IGS, and what influences their perception.

In this study, we explore the potential of IGS as a supplementary urban green space in contributing to well-being in the urban environment given the spatial and financial constraints of Asian cities with a high population density, as represented by the case of Ichikawa City, Japan. To consider and evaluate IGS as supplementary green space in cities, we focus here on its perception by residents. Moreover, since IGS is not an officially recognized space as a formal classification category for either conservation or recreation, it may be difficult for residents to perceive IGS as a UGS. We hypothesize that their attitude towards green space is not based on formal education, but rather formed through experience and influences in real life. Therefore, to explore the issues, this paper seeks to contemplate the understanding of IGS against the background of residents' perception of existing green spaces, such as urban parks.

We focused on the following research questions: (1) What are the merits of IGS that residents perceive and why are they reluctant to use IGS; (2) how does IGS perception differ depending on UGS experience; (3) how do residents perceive IGS depending on their residential environment; and (4) what is the difference and relation between residents' attitudes toward urban nature, including UGS, and IGS perception?

2. Materials and Methods

2.1. Study Site

Our study site was Ichikawa (57.10 km^2 with 482,544 inhabitants), located in the Chiba Prefecture, Japan (Figure 1). This city has been formed while being strongly influenced by outer Tokyo. There have been three waves of rapid population inflows without prior establishment of urban infrastructure due to its location close to the capital of Japan. Land readjustment projects and railway construction projects have created high density urban districts. Currently, Ichikawa consists of more than 70% urbanized areas, including residential, commerce, and industrial districts, and about 30% (29.24%) of urbanization control area intended to constrain periurban sprawl.

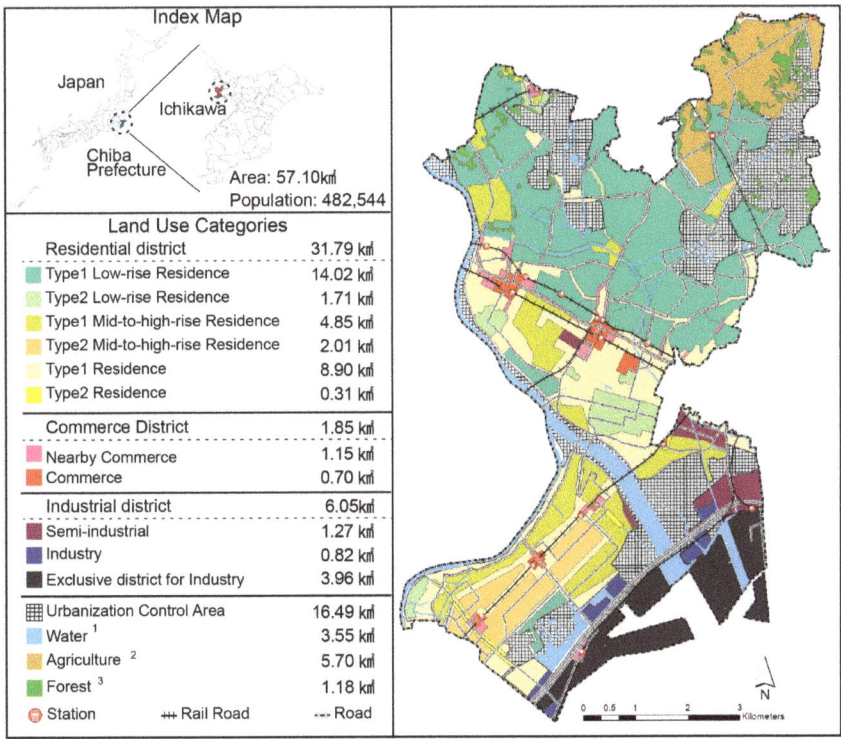

Figure 1. Location of and land use categories in Ichikawa, Japan. [1] Some of the total water areas overlap with urbanization control areas. [2] The agricultural district is included in the controlled urbanization district. [3] Forest area exists not only in agricultural districts, but also residence districts and controlled urbanization districts.

Acquisition of public land by the city is not easy because districts have formed dense urban areas of narrow roads and their land price has risen [30]. Since most citizens migrated from outside the city, the general sense of community attachment is low. This phenomenon influenced the city government to attempt addressing it through urban plans and creating green spaces. Ichikawa government has implemented several town plans for improving residents' quality of life since the year, 2000 [30,31]. According to the Green Master Plan of Ichikawa, the government aimed to improve green space from 2003 to 2025 in three steps, using green space per capita (m^2/person) as an indicator. The indicator at the time they declared the plan was 2.70 m^2, and the next goal was set at 3.85 m^2 for 2015 before the final goal of 4.73 m^2 per capita by 2020 [32]. However, the city only had 3.43 m^2 per capita as of 2016, and it seems unlikely that it is possible to provide residents with equal opportunity to use green space according to the Urban Park Act of Japan, which recommends 10.0 m^2 per capita.

2.2. Data Collection, IGS Typology, and Data Analysis

We conducted a survey targeting residents using a mail-back questionnaire distributed around the sample sites (Figure 2a) of an existing grid that was set up for a previous field survey of IGS distribution. Sampling kits were allocated at 20 per sample site, and a total of 3700 kits were distributed, except in the non-resident areas. If there were not enough residences in the sample site, we extended the distribution scope using a buffer as 50 m or 100 m focusing on the sites. The number of replies per site was from 1 to 8, with an average of 3.29 responses (Figure 2b).

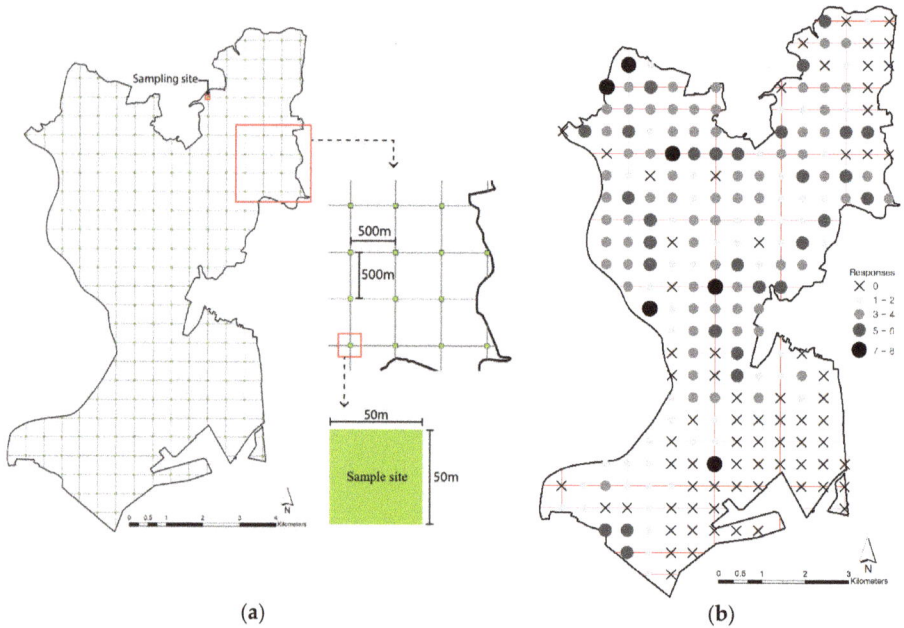

(a)   (b)

Figure 2. Sampling strategy and number of responses: (**a**) Distribution of survey sites across Ichikawa; (**b**) number of responses per sample site.

Before creating the survey instrument, we conducted a pilot workshop on IGS with 70 undergraduate students of agricultural science and landscape architecture. We discussed the merits IGS is considered to have and reasons why one may be reluctant to use it. Results were used to create the questionnaire. The questionnaire contains questions on general characteristics of the respondents, the merit of IGS, potential reasons for their reluctance to use IGS, and on respondents' attitudes toward urban green space. To ensure the contents of the questionnaire were easy to understand and answer for residents without a relevant professional or academic background, grammar and wording were revised by seven native non-specialist Japanese speakers. To capture the full variety of IGS in Ichikawa, we extended the IGS typology by adding 'parking lot verges' and 'unimproved land' to the typology used in previous work [28] (Table 1). Additionally, we provided photos of the revised IGS types in our questionnaire sheet to allow residents to visually identify what IGS looks like (Figure 3). We lowered the color saturation of the non-IGS area in the photos to make it easier for residents to notice IGS in the images provided.

Table 1. Description of the nine types of informal urban green space (IGS).

IGS		Description (Non-Exclusive Criteria)
Vacant lots	Profile	Space left unused after its previous use ends. The site may be empty, or the infrastructure of the building's frame or debris from the building remain. Former use was primarily housing, but it is now unused and neglected.
	Vegetation	The type of vegetation differs depending on the status of the management of the space and the period left from the time when the original usage ends. The pattern of vegetation ranges from well-trimmed grass to small-scale bushes where succession has progressed to some extent.
	Maintenance and Access	Management is carried out irregularly with minimum maintenance, such as mowing the lawn. However, there are many places where management is not done for a long time. Access is restricted by fences or signs to protect private property, but some are open space.
Street verges	Profile	Mainly located on the perimeter of a driveway or pedestrian road.
	Vegetation	The pattern of vegetation consists mainly of herbaceous plants, which are dominated by spontaneous vegetation. Vegetation begins to spread linearly among heterogeneous pavement materials.
	Maintenance and Access	It is usually managed irregularly by the government and contractors rather than individuals, and plant cutting activities are often carried out in response to residents' complaints. There are no elements, such as fences or signs, to restrict access, and the accessibility depends on where they are located.
Water verges	Profile	Formed by vegetation within 10 m from the water body. The type of the area includes all sections where water flows, such as river, canal, stream, waterway, and watersheds.
	Vegetation	Vegetation communities can be directly tied to water bodies, or they grow on land within 10 m of the water bodies. Unlike intended planting patterns for a recreational purpose, such as a waterside park, these are spontaneous vegetation communities.
	Maintenance and Access	Government agencies usually conduct management. For the non-waterfront parks, the management activities focus on monitoring for disaster prevention or the quality of water. Most of them are difficult to access to the water center due to fencing or signs.
Gaps	Profile	Vegetated space formed between structures. The spaces of structures include between walls, between fences, and between remaining building structures.
	Vegetation	Most of the space in the gap is covered with herbaceous plants.
	Maintenance and Access	Space management is carried out on an irregular basis, if at all. Most of the management activities are cutting overgrown plants and disposing of garbage.
Brownfields	Profile	Space where the existing infrastructure has remained as all or a part after the end of the previous use and not used at present. The previous uses of space are mainly by the light industry or commerce, not housing.
	Vegetation	Vegetation is spontaneously scattered in an atypical shape influenced by existing planting space, cracks, and heaps of dirt.
	Maintenance and Access	Largely neglected space whose original use has been terminated and the access of the public is controlled. Vegetation and spaces are rarely managed.
Unimproved lands	Profile	Empty land without infrastructure, such as electricity and sewage facilities; has the potential for development at any point in time. It is located in periurban areas rather than the central portion of the city, such as the 'Urban Control District'.
	Vegetation	Most of the vegetation is composed of spontaneous herbaceous plants, but, in some cases, a small number of trees have been planted intentionally by a landowner.
	Maintenance and Access	Since the site is not currently being used for any other purpose, systematic and regular management does not occur. In the case of some places that are located away from the center of the city, vegetation succession has progressed and forms a meadow because management has not been carried out for a long time.
Parking lot verges	Profile	Site representing a secondary use of a 'vacant lot' rather than a planned place for parking. The site features minimal land maintenance and separation of parking spaces. Distinct from an automated parking lot operated by a professional enterprise.
	Vegetation	Vegetation is clustered linearly around the edge of the parking lot and is dominated by spontaneous herbaceous plants, and not by intentional plantings.
	Maintenance and Access	Minimal maintenance is performed regularly for the function of the parking lot. Vegetation communities formed on the edges are often removed due to parking lot users' complaints.
Railroad verges	Profile	Space with vegetation adjacent within 10 m of railway tracks.
	Vegetation	Vegetation forms linearly along the track or forms communities around a station.
	Maintenance and Access	For reasons of safety, direct public access is strictly controlled. Removal of plants or use of herbicides is carried out irregularly.
Overgrown structures	Profile	Space where plant communities cover artificial structures and often grow vertically.
	Vegetation	These spaces are predominantly dominated by vines. In the case of public buildings or structures with no safety concerns, there are sometimes intentional plant patterns to improve the thermal environment.
	Maintenance and Access	There may be differences in public accessibility depending on the type and location of the structure. If structural safety is to be maintained, plants are regularly removed, and public access is blocked.

Figure 3. Nine types of IGS in Ichikawa.

We compared the differences and characteristics of the perceptions of IGS from the two perspectives of IGS eight merits (ME) and eight reasons for reluctance to use IGS (RE): (ME.1) IGS makes urban landscape beautiful; (ME.2) IGS can make me feel nature in an urban area; (ME.3) IGS is easy to access because it is close to where I live; (ME.4) it is possible to use IGS freely in many ways; (ME.5) IGS can be a place where children can play; (ME.6) IGS can be a habitat for living things; (ME.7) IGS has the effect of suppressing dust; (ME.8) IGS can be useful for air purification; (RE.1) I'm concerned about the conflict with the landowner of the site; (RE.2) signs or fences make it difficult to get into the site; (RE.3) risk of injury; (RE.4) there is a lot of trash inside; (RE.5) it seems to be polluted; (RE.6) it is not managed for use; (RE.7) it is too small or narrow to use; and (RE.8) it may be either developed or disappear someday. We therefore used 'ME' and 'RE' as dependent variables and used as independent variables the general attributes of respondents, experience with UGS, the relationship between surrounding greenery and residence environment, and attitude towards UGS. We organized the attitude of residents toward urban green space (AT) into ten categories based on the pilot workshop: (AT.1) I cherish the urban nature with plants and animals; (AT.2) UGS makes my everyday life environment healthy; (AT.3) it is important to coexist with plants, animals, and humans in an urban environment; (AT.4) I'm willing to participate as a volunteer to conserve nature; (AT.5) I'm willing to arrange a time for conserving nature; (AT.6) I'm willing to pay some money to conserve nature; (AT.7) I've known plants, animals, and insects that are often observed in or near my area; (AT.8) I can feel the community attachment from plants, animals, and insects that are often observed in or near my area; (AT.9) the neighborhood green space should be managed; and (AT.10) the neighborhood green space should be convenient. For the variables for each section, the values for asymmetry and kurtosis were considered acceptable between −2 and +2 to prove normal univariate distribution [33–35], but we found that some variables did not meet normality. Therefore, we used the Mann-Whitney U test as a nonparametric test method to compare differences in IGS perception, and the Chi-Square test (X^2 test) to analyze observations for statistically significant results. We also conducted a factor analysis to reduce and interpret the 10 attitudes toward the urban nature including UGS into useful factors. The reliability of the variables by factor analysis was tested using the Cronbach Alpha test of Kaiser-Meyer-Olkin measure of sampling adequacy (KMO). Logistic regression analysis was used to measure which factors can be classified into IGS perception using the forward conditional method after identifying the correlation between the IGS perceptions and attitudes. We verified the fitness of the logit model by the Hosmer-Lemeshow test (Figure 4). To statistically analyze and chart the questionnaire, we used Excel 2016 and IBM SPSS (version 25) software.

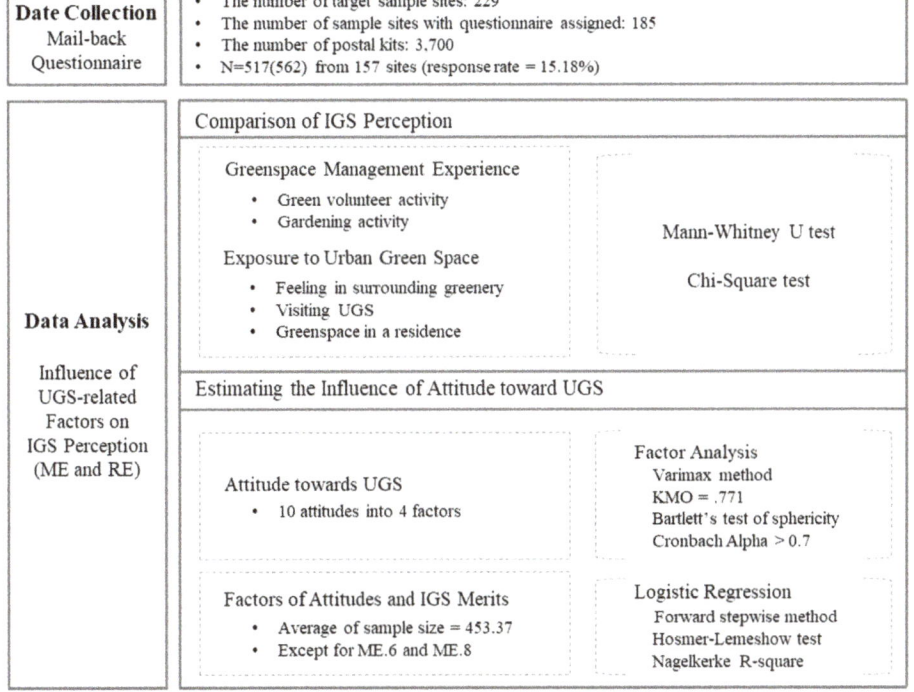

Figure 4. Research workflow.

3. Results

3.1. Demographic Composition and Sample Characteristics

Of the 229 planned distribution sites, 44 sites were excluded because they contained either no-residence or were inaccessible. We thus distributed the survey kits to 185 sites (total 3700 kits) and received 562 responses from 157 sites in about two months (response rate = 15.18%). Some enthusiastic respondents sent comments on IGS and its survey contents using memos and letters. Respondents were 59.6% female and 41.4% male, and respondents over 50 years accounted for 60% of respondents (Table 2). To understand residents' stance toward IGS perception in the context of greenery in their everyday life, we asked questions about three topics: UGS-related experience, greenery contact, and attitude towards urban nature. Respondents had little experience, such as being a green volunteer in public spaces (urban parks, protected forests etc.), but more than 60% of all respondents had experience with private spaces, such as home gardens, verandas, or allotments. Those who had never visited surrounding green space were 10% higher than those who went there every day. For contact with greenery within the residential environment, about 80% of respondents could access green space within their residential range in the form of a home garden or shared green space. Moreover, residents who thought that there was plenty of green space around their living environment were about 10% higher than those who felt green space lacking.

Table 2. Respondents composition (n = 517).

Respondents Composition		Total	(%)
Gender	Male	214	41.4
	Female	303	58.6
Age	20–29	27	5.2
	30–39	56	10.8
	40–49	105	20.3
	50–59	98	19.0
	60–69	108	20.9
	Over 70	123	23.8
Children in family	No	374	72.3
	Yes	143	27.7
Employment status	Unemployed or retired	218	42.2
	Employed	299	57.8
Public experience [1]	No	422	81.6
	Yes	95	18.4
	Mean participation frequency: 23.12 (minimum value = 1, maximum value = 1000, SD = 109.077, n = 86)		
Individual experience [2]	Never	93	18.0
	Sometimes	88	17.0
	Ongoing	336	65.0
Frequency of visiting green space	Never	155	30.0
	1~3 times a year	93	18.0
	1~3 times a month	94	18.2
	1~3 times a week	70	13.5
	everyday	105	20.3
Housing type	Detached house with green space	300	58.0
	Detached house without green space	60	11.6
	Apartment with shared green space	105	20.3
	Apartment without shared green space	52	10.1
Recognition of the quantity of surrounding greenery	Strongly lacking	27	5.2
	Lacking	133	25.7
	Moderate	136	26.3
	Considerable	171	33.1
	Plenty	50	9.7

[1] Public experience here refers to green space conservation activity like volunteering for improving the public environment in parks, forests, and rivers. The main activities are tree planting, weeding, cleaning, agricultural experience, observing fauna and flora, and monitoring introduced species. [2] Individual experience here refers to gardening activity to grow and manage plants in the home garden or veranda. This activity focuses more on individual satisfaction than on the improvement of the public environment.

3.2. Merits of IGS and Reasons for Reluctance to Use IGS

Before exploring how IGS perception was influenced by residents' green space contact in daily life, we asked about the overall merits (ME) of IGS that residents were aware of and why they were reluctant (RE) to use it. When comparing perceived merits and reluctance, most of the residents more strongly felt the benefits of IGS than a reluctance to use it (Figures 5 and 6). Residents valued IGS aesthetically (ME.1 and ME.2) and its environmental functions (ME.6 to ME.8) higher than its recreational aspects (ME.3 to ME.5). There was no difference in perception of IGS merits according to respondents' general characteristics, such as gender, having children in the family, and employment status. However, age was related to ME.3. As the age range of the respondents increased, they recognized that having IGS close to where they reside as an advantage (X^2 = 52.141, sig(p) = 0.000).

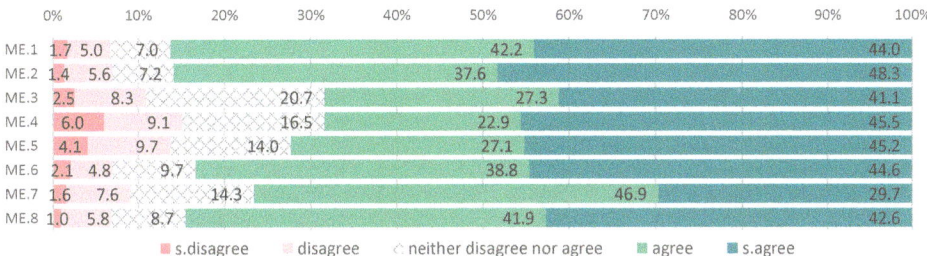

Figure 5. The merit of IGS. (ME.1) IGS makes urban landscape beautiful; (ME.2) IGS can make me feel nature in an urban area; (ME.3) IGS is easy to access because it is close to where I live; (ME.4) it is possible to use IGS freely in many ways; (ME.5) IGS can be a place where children can play; (ME.6) IGS can be a habitat for living things; (ME.7) IGS has the effect of suppressing dust; and (ME.8) IGS can be useful for air purification.

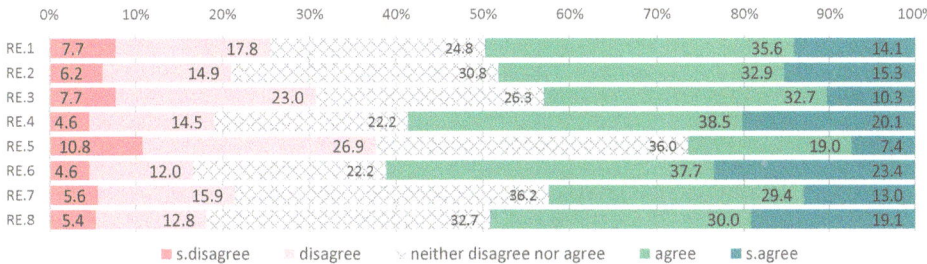

Figure 6. Reasons for reluctance to use IGS. (RE.1) I'm concerned about the conflict with the landowner of the site; (RE.2) signs or fences make it difficult to get into the site; (RE.3) risk of injury; (RE.4) there is a lot of trash inside; (RE.5) it seems to be polluted; (RE.6) it is not managed for use; (RE.7) it is too small or narrow to use; and (RE.8) it may be either developed or disappear someday.

When asked about reasons for their reluctance to use IGS, respondents were more sensitive to the current non-managed status (RE.4 and RE.6) than concerns that might arise when actually using it (RE.1 to RE.3). They perceived IGS as an unmanaged space, but they did not agree that it was dirty or contaminated (RE.5). Respondents' general characteristics, such as age and childcare, influenced responses to RE.6. Young and child-care respondents were more aware of IGS as an unmanaged space. Respondents from teens to those up to 49 years old more strongly agreed on 'RE.6' than respondents over 50 years of age; younger respondents agreed to RE.6, with an average of 71.6%, while the over 50 s agreed on it, with an average 55.2% (X^2 = 22.835, sig(p) = 0.029). In addition, respondents who were raising children strongly agreed on 'RE.6' with 69.9% compared to those who are not (X^2 = 7.142, sig(p) = 0.028).

3.3. Influence of UGS-Related Factors on IGS Perception

3.3.1. Greenspace Management Experience

We sorted the greenspace management experiences into two groups based on where the experiences took place. Green volunteer activity refers to conservation activity in public areas, such as parks, forests, and rivers, etc. This activity involves tree planting, weeding, cleaning, observing fauna and flora, and monitoring introduced species. The effect of these activities may encourage participants to be considerate of the public environment. In contrast, gardening activity refers to horticultural activities for self-satisfaction and improvement of personal living environments. This activity takes place in private spaces, such as private home gardens, verandas, and allotments. The gardening performers seek individual aesthetic and therapeutic effects for mental health and/or consumption of food [36]. We used the Mann-Whitney U test to compare how having or not having experience in either the public or private space affected the perception of IGS merits and reasons for reluctance to use IGS. We reclassified the existing three items of the frequency of gardening activity into two items: 'No' and 'yes'. Mean rank of the Mann-Whitney U test indicated that people who have experienced UGS management had a more favorable position toward IGS merits. Besides, the result of the experiment demonstrated that people who had experienced gardening activities were less reluctant to use IGS, as shown by the significant difference in responses between the variables for four MEs and two REs (Table 3). Volunteer experience was associated with differences between the variables in the response about environmentally functional aspects of IGS, but no significant difference was found regarding a reluctance to use IGS.

Table 3. Mann-Whitney U test result of the urban green space (UGS) experience (n = 517).

Green Volunteer Activity		ME.6		ME.7		ME.8	
Mean Rank	No (n = 422)	254.08		253.89		253.14	
	Yes (n = 95)	256.94		281.68		285.05	
Mann-Whitney U		17,967.000		17,890.000		17,570.000	
Z		−2.428 *		−2.215 *		−2.993 **	
Gardening Activity		**ME.3**	**ME.6**	**ME.7**	**ME.8**	**RE.4**	**RE.6**
Mean Rank	No (n = 93)	219.44	235.29	227.58	229.48	288.06	291.84
	Yes (n = 424)	267.68	264.20	265.89	265.47	252.63	251.80
Mann-Whitney U		16,036.500	17,511.000	16,794.000	16,971.000	17,013.000	16,661.500
Z		−3.442 **	−2.598 **	−3.028 **	3.347 **	−2.348 *	−2.693 **

* $p < 0.05$, ** $p < 0.01$ (ME.3) IGS is easy to access because it is close to where I live; (ME.6) IGS can be a habitat for living things; (ME.7) IGS has the effect of suppressing dust; (ME.8) IGS can be useful for air purification; (RE.4) there is a lot of trash inside; and (RE.6) it is not managed for use.

Based on the differences in variables identified above, we visualized Chi-Square (X^2-test) test results to compare the observed counts (Figure 7). Gardening activity in the X^2-test was compared with the existing three items based on the frequency of gardening experiences as 'Never', 'Sometimes', and 'Ongoing'. Since about 81% of all respondents had no public green-related volunteer experience, the 'No (no experience)' proportion was relatively high in responses to all ME. In this pattern of responses, however, we found a change in the proportion on each answer from 'disagree', 'neither', and 'agree' from ME. Although there are no statistically significant differences on ME.6 and ME.7 in the X^2-test, the proportion of respondents agreeing on the air purification merits of IGS (ME.8) was higher in those with volunteer experience. Since 65% of all respondents are doing gardening every day as well, the proportion of experienced respondents is high. Those who do every day horticultural activities account for a higher rate of positive ME perceptions, while those who have never or rarely done horticultural activity had a more negative stance. The proportion of people who do gardening activity daily was 50.8% on average in those with a negative stance towards IGS merits and 69.8% in those with a positive view. The difference of opinion according to whether respondents engaged in garden activity was largest in ME.7. People who had never or rarely experienced gardening activity were more skeptical of IGS merits and agreed more strongly with reasons to be reluctant to use IGS. The responses to 'RE.4 and 'RE.6' showed statistically significant differences.

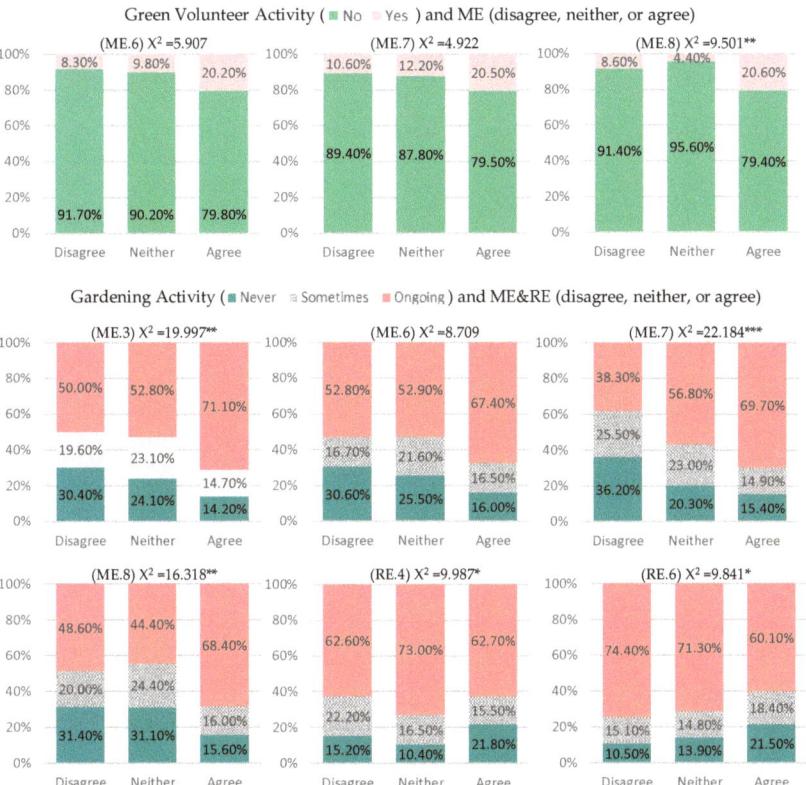

Figure 7. X^2-test between greenspace management experience and ME&RE; * $p < 0.05$, ** $p < 0.01$, *** $p < 0.001$ (ME.3) IGS is easy to access because it is close to where I live; (ME.6) IGS can be a habitat for living things; (ME.7) IGS has the effect of suppressing dust; (ME.8) IGS can be useful for air purification; (RE.4) there is a lot of trash inside; and (RE.6) it is not managed for use.

3.3.2. Exposure to Urban Green Space

We categorized environmental contact with green spaces into three types: First, how much green space do residents perceive in their living surroundings? Second, what kind of green space is connected to residents in their residential environment? Third, how often do residents use UGS? Asked how much green spaces residents perceive in their surrounding environment, 221 respondents (42.8%) responded that green spaces are abundant, while 160 (31.2%) answered that green spaces are lacking. Four hundred and five respondents were living in housing with green space, of which 72.07% of them could access green space by a home garden from the house, and 25.93% shared green space within an apartment housing. The proportion of people who do not use UGS at all was about 2% higher than the proportion of people visiting UGS every day. We divided the responses regarding environment toward surrounding greenery into two groups: Low and high green space exposure. In these groups, we excluded neutral responses and compared the perception of 'ME and RE' of IGS. Table 4 shows significant values for differences in IGS perception for each independent variable. The group with high amounts of green space exposure had a more positive stance toward IGS merits. Moreover, residents who could access green space from their home garden in the residential environment showed a higher position on 'ME.7' than people who could access green space as a shared form. The group with low green space exposure agreed more strongly with reasons for being reluctant to use IGS.

Table 4. Mann-Whitney U test results of exposure to urban green space.

Feeling in Surrounding Greenery		ME.3	ME.6	RE.1	RE.5	RE.7	RE.8
Mean Rank	Lacking (n = 160)	150.50	180.15	204.18	208.01	215.87	203.43
	Abundant (n = 221)	220.32	198.86	181.46	178.68	173.00	182.00
Mann-Whitney U		11,200.000	15,944.000	15,572.000	14,958.000	13,701.500	15,691.500
Z		−7.427 ***	−2.602 **	−2.154 *	−2.732 **	−4.028 ***	−2.047 *
Visiting UGS		ME.3	ME.4	ME.6	RE.4	RE.5	RE.6
Mean Rank	Never (n = 155)	144.43	148.03	152.92	176.04	175.94	176.25
	Frequently (n = 175)	184.16	180.97	176.64	156.17	156.26	155.98
Mann-Whitney U		10,296.500	10,855.000	11,612.500	11,929.500	11,945.000	11,896.000
Z		−4.753 ***	−3.799 ***	−3.381 **	−2.115 *	−1.994 *	−2.200 *
Green Space in a Residence		ME.3	ME.7	Green Space in a Residence			ME.7
Mean Rank	Nothing (n = 112)	222.27	238.44	Mean	Home garden (n = 300)		211.74
	Contacting (n = 405)	269.16	264.69	Rank	Shared Green Space (n = 105)		178.03
Mann-Whitney U		18,566.000	20,377.000	Mann-Whitney U			13,128.500
Z		−3.588 ***	−2.225 *	Z			−3.547 ***

* $p < 0.05$, ** $p < 0.01$, *** $p < 0.001$ (ME.3) IGS is easy to access because it is close to where I live; (ME.4) it is possible to use IGS freely in many ways; (ME.6) IGS can be a habitat for living things; (ME.7) IGS has the effect of suppressing dust; (RE.1) I'm concerned about the conflict with the landowner of the site; (RE.4) there is a lot of trash inside; (RE.5) it seems to be polluted; (RE.6) it is not managed for use; (RE.7) it is too small or narrow to use; and (RE.8) it may be either developed or disappear someday.

All independent variables had significant influence on 'ME.3'. We have visualized a summary of the respondents' groups' cases regarding contact with the green environment for 'ME.3' among the IGS perception variables (Figure 8). In the case of the respondents who had relatively less access to the green environment in their residential area than home garden owners, the perception of 'ME.3' significantly increased with more UGS visits. In other words, residents who did not exclusively use green space within their dwellings had a notably higher perception of IGS proximity according to the frequency of UGS visits (Figure 8a). For the respondents who had no green space attached to their dwellings, agreement with 'ME.3' increased with the greenery they perceived around their residential area. There was a significant difference in the perception of 'ME.3' between those perceived to lack green space and those perceived as moderate (Figure 8b).

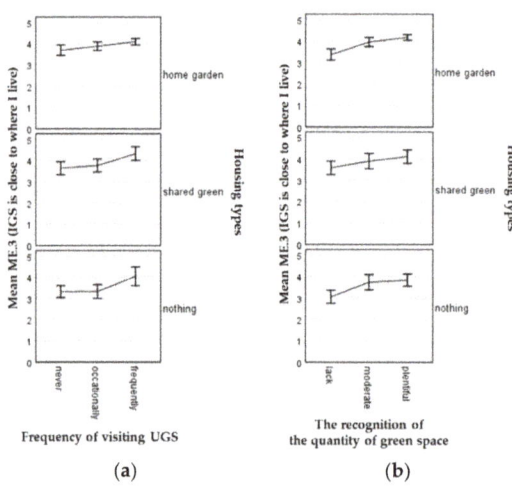

(a)   (b)

Figure 8. Exploring independent variable effects on 'ME.3' (Error bars: 95% CI). (a) Frequency of visiting UGS for different housing types; (b) perceived quantity of green space for different housing types.

3.3.3. Attitude towards Urban Green Space

Although IGS is not an officially recognized green space, such as an urban park, we hypothesize that to meet recreational or aesthetic needs of users, even in liminal spaces, naturally occurring vegetation may provide the potential to supplement UGS. Therefore, we tested how perception of IGS was affected by respondents' general attitude toward UGS and the urban environment. We asked residents ten questions about their attitude (AT) towards not only UGS, but also the urban environment, and identified factors with a factor analysis to investigate variable relationships for mixed concepts using varimax rotation (Figure 9).

(AT. F.1) Willingness to participate in conservation activities of urban green space and urban nature.
Cronbach Alpha: 0.845

(AT. F.2) Prospect for coexisting to human and urban nature.
Cronbach Alpha: 0.819

(AT. F.3) Community attachment through neighboring nature.
Cronbach Alpha: 0.786

(AT. F.4) Usability and necessity of management of the neighboring green space
Cronbach Alpha: 0.765

Figure 9. Residents' attitude to UGS and urban nature. (AT.1) I cherish the urban nature with plants and animals; (AT.2) UGS makes my everyday life environment healthy; (AT.3) it is important to coexist with plants, animals, and humans in an urban environment; (AT.4) I'm willing to participate as a volunteer to conserve nature (AT.5) I'm willing to arrange a time for conserving nature; (AT.6) I'm willing to pay some money to conserve nature; (AT.7) I've known plants, animals, and insects that are often observed in or near my area; (AT.8) I can feel the community attachment from plants, animals, and insects that are often observed in or near my area; (AT.9) the neighborhood green space should be managed; and (AT.10) the neighborhood green space should be convenient.

The conducted four valuable factors were derived with a Kaiser-Meyer-Olkin Measure of Sampling Adequacy (KMO) value of 0.771 and Bartlett's Test of Sphericity (Approx. Chi-Square: 2241.887, df:45, Sig(p): 0.000), and the identified factors were tested by calculating their Cronbach Alpha. Even though respondents were not overly confident in their knowledge of UGS, harmony between non-human and human life in urban areas was considered important by respondents (AT.F.2). However, their attachment to close-by nature was weaker than their belief in the value of coexistence with nature (AT.F.3). Residents were generally in favor of participation in conservation activities of UGS or urban environment, but not in a very active way (AT.F.1). Regarding the usability and necessity of management of neighborhood green space, opinions were distributed relatively evenly (AT.F.4). Of the derived four factors, AT.F.1 and AT.F.2 were correlated with all items of ME. AT.F.3 was correlated with all 'ME' items except ME.5, and AT.F.4 was related to the recreation potential of IGS, ME.4, and ME.5. Concerning RE, there was a correlation with a few variables, but none with most. AT.F.1 and AT.F.2 correlated with RE.5 and RE.6, which are related to the management status of IGS. AT.F.3, which referred to the local attachment, was correlated with RE.1, RE.6, and RE.7. Finally, AT.F.4 correlated with RE.3 and RE.5, which implies a negative perception of non-management.

We established the correlation with ME and RE as the dependent variables by conducting a logistic regression. We rearranged the group of the dependent variables into binary forms of 'agree' and 'disagree' for IGS's ME and RE and excluded the neutral opinion, 'undecided'. As a result, the size of the samples corresponding to each dependent variable was less than the overall sample of this study (n = 517). The sample size for each variable ranged between 409 to 481, with an average of 453.37. The correct percentage of classifying the attitude factors for ME variables was high, ranging from 81.8% to 92.7%. On the other hand, the correct classification of attitude factors for RE variables was 65.66% on average. In this study, the explanatory power Nagelkerke R Square for the regression model for the ME variables was 0.160 on average (Min: 0.103, Max: 0.273), and for the RE variables the average was 0.031. Thus, we performed the logistic regression on the ME variables, excluding the RE variables with low classification accuracy and explanatory power. Among the results of the logistic regression toward ME, we also excluded ME.6 and ME.8, where the fitness of the logit model by the Hosmer-Lemeshow test was not established (Table 5). We found that the factors for ME were all significant ($p < 0.05$). For the odds ratio (Exp(B)) value, which can identify the change of the probability of 'ME' recognition as each attitude factor variable increases, the AT.F.1 variable [Exp(B) = 3.330] corresponding to the ME.1 dependent variable was the highest. The willingness to participate in the conservation activities of an urban nature (AT.F.1) was included as an element increasing the probability of the perception in all ME dependent variables. AT.F.4, the usability and necessity of management of the neighboring green space, has been included as a recognition element of ME.4 and ME.4, the recreational aspect of IGS.

Table 5. The results of the logistic regression.

Dependent Variable	Independent Variables	B	S.E.	Wald	df	Sig.	Exp(B)
ME.1 (beauty) [1]	AT.F.1	1.203	0.199	36.603	1	0.000	3.330
	AT.F.2	0.529	0.166	10.206	1	0.001	1.697
	AT.F.3	0.432	0.177	5.980	1	0.014	1.540
	Constant	3.227	0.276	139.501	1	0.000	25.199
Classification percentage = 92.7%, Nagelkerke R^2 = 0.273, Hosmer and Lemeshow test Chi-square = 5.105 (df = 8, Sig(p) = 0.746)							
ME.2 (nature)	AT.F.1	0.793	0.167	22.576	1	0.000	2.210
	Constant	2.764	0.209	174.147	1	0.000	15.860
Classification percentage = 92.5%, Nagelkerke R^2 = 0.114, Hosmer and Lemeshow test Chi-square = 7.174 (df = 8, Sig(p) = 0.518)							
ME.3 (close)	AT.F.1	0.397	0.140	7.982	1	0.005	1.487
	AT.F.2	0.479	0.137	12.133	1	0.000	1.614
	AT.F.3	0.294	0.142	4.283	1	0.038	1.341
	Constant	1.936	0.157	151.227	1	0.000	6.930
Classification percentage = 86.3%, Nagelkerke R^2 = 0.103 Hosmer and Lemeshow test Chi-square = 9.008 (df = 8, Sig(p) = 0.342)							
ME.4 (activity)	AT.F.1	0.761	0.137	31.024	1	0.000	2.140
	AT.F.2	0.485	0.135	12.942	1	0.000	1.623
	AT.F.3	0.447	0.132	11.480	1	0.001	1.563
	AT.F.4	0.342	0.135	6.429	1	0.011	1.408
	Constant	1.747	0.151	134.029	1	0.000	5.738
Classification percentage = 81.9%, Nagelkerke R^2 = 0.222 Hosmer and Lemeshow test Chi-square = 3.670 (df = 8, Sig(p) = 0.886)							

Table 5. *Cont.*

Dependent Variable	Independent Variables	B	S.E.	Wald	df	Sig.	Exp(B)
	AT.F.1	0.510	0.136	14.000	1	0.000	1.666
ME.5	AT.F.2	0.414	0.124	11.078	1	0.001	1.153
(children)	AT.F.4	0.374	.0132	7.986	1	0.005	1.453
	Constant	1.803	0.145	154.854	1	0.000	6.070

Classification percentage = 84.0%, Nagelkerke R^2 = 0.127
Hosmer and Lemeshow test Chi-square = 6.370 (df = 8, Sig(p) = 0.606)

Dependent Variable	Independent Variables	B	S.E.	Wald	df	Sig.	Exp(B)
	AT.F.1	0.537	0.154	12.093	1	0.001	1.711
ME.7	AT.F.2	0.415	0.142	8.565	1	0.003	1.151
(dust)	AT.F.3	0.619	0.157	15.468	1	0.000	1.857
	Constant	2.408	0.190	160.820	1	0.000	11.115

Classification percentage = 89.4%, Nagelkerke R^2 = 0.160
Hosmer and Lemeshow test Chi-square = 5.133 (df = 8, Sig(p) = 0.743)

[1] The words in parentheses are keywords that can describe each dependent variable. (ME.1) IGS makes urban landscape beautiful; (ME.2) IGS can make me feel nature in urban area; (ME.3) IGS is easy to access because it is close to where I live; (ME.4) it is possible to use IGS freely in many ways; (ME.5) IGS can be a place where children can play; and (ME.7) IGS has the effect of suppressing dust.

4. Discussion

This study was conducted to consider IGS as a supplementary urban green space in response to the physical and financial constraints in green provisioning in contemporary urban areas. Overall, we suggest our findings support the view that IGS has potential to supplement UGS in Ichikawa. However, IGS is not officially designated or recognized by the government or landowner for a recreational or protective purpose [28]. Therefore, it may be difficult for residents to perceive IGS as a stance equivalent to existing UGS, such as urban parks. Understanding these issues, we investigated the perception of IGS from the point of existing UGS that is already familiar to residents. In this context, we examined residents' IGS perception as influenced by their experience, green space exposure, and attitude towards UGS. We discuss the implications of our findings in more detail in the following section.

4.1. The More Favorable Towards UGS, The More Favorable Towards IGS

In general, familiar objects are recognized categorically, and perceptual similarities are closely related to perceived objects [37]. Respondents who have a close relationship with UGS in their daily lives tend to have a favorable perception toward IGS, even if IGS is not designated by the government or landowner for recreational use. Overall, respondents took a favorable stance to IGS, but there were differences in their positions according to the frequency of their exposure to UGS and their experience of greening-related activities. Respondents who actively engaged with the environment, such as green volunteering and gardening, tended to perceive IGS as a medium that may improve environmental issues in urban areas, for instance, air purification and dust suppression. They see the possibility to improve the surrounding environment due to the spontaneous vegetation within IGS. Respondents who use UGS more also recognized IGS more as a spatial element which people can use and act. In contrast, the respondents with little experience about UGS took a skeptical stance to using IGS. Respondents who have no active UGS experience, such as visiting and managing greenspace, recognized IGS as an unmanaged and neglected space (see Tables 3 and 4). The respondents who were not satisfied with the quantity of UGS in their surroundings felt more uncertain about using IGS and were concerned with the landowners (see Table 4). Therefore, considering that favorable perception toward IGS is linked to the degree of UGS experience, one cause for this may be the perceptual similarity between UGS and IGS. This relationship between green space experience and perception of IGS suggests that urban green space can be supplemented, but more so in areas where a certain level of UGS are already provided and for residents who already use UGS. In contrast, these findings suggest that unlocking the potential of IGS for recreation for areas with very little UGS and residents unfamiliar with UGS may be challenging.

4.2. IGS: Located Close By and Easy to Access

Many studies show that green space is an essential component of urban space as an open space for improving the sustainability of the urban environment and the health of residents. In the context of these issues, contemporary researchers are concerned about accessibility to urban green spaces as access is linked to improvements in residents' health and social well-being [38–40]. In previous research on the recognition of green space with IGS [26,41], accessibility from home was perceived as an important reason why participants used IGS. Our survey results show that the accessibility aspects of IGS are highly influenced by UGS experience. Those with UGS experience perceived that IGS that is located near their residence as an advantage. In contrast, residents who lack access to green space in their housing and are less satisfied with their surrounding green environments had a lower perception of IGS proximity. This is important because accessibility and quantity of green space are linked to maintaining the well-being of residents of the neighborhood, particularly housewives, the elderly, and those who are socially vulnerable [42,43]. Improving accessibility to open space, including green space, has been shown to play an important role for the elderly in encouraging their physical activity and quality of life [44,45]. Moreover, living nearby a relatively comfortable and walkable green space was correlated with a lower mortality risk for older people [46]. In this context, IGS proximity is of particular interest when taking access to greenspace for aging residents into account. Most of the respondents in our study were of a high age. This demographic composition reflects the current situation in Japan, which has entered a super-aged society. In fact, the proportion of the aging population in Ichikawa was 23.8% by 2015. In our study, older adults took part more frequently in green activities and responded more to the surrounding green environment. The beneficial effects from green space are more pronounced in elderly and housewives who rely more on the local living environment [47]. According to a recent study, about 67% of adults over the age of 60 spent 8.5 h indoors on a sedentary basis [48]. Therefore, given the age groups surveyed and the IGS accessibility they perceive, IGS may serve as an element that not only encourages outside activity and promotes physical health, but also promotes social cohesion and a community for older adults, for whom social isolation has been linked with increased mortality [49,50].

4.3. IGS and Participatory Aspects

Green spaces managed by local residents enhance the local biodiversity and ecosystem services production and encourage user participation [51]. While the structural complexity and intensity of management influences the basis of participation, appropriate participatory management provides an opportunity to improve participants' health with physical activities through the management of the site [52,53]. In our study of the residents' attitudes toward UGS, they were significantly less willing to conserve urban nature than to agree that urban nature and human beings must coexist. Although the level of willingness to participate was overall rather low in this study, the results confirm findings of previous research on willingness to participate in IGS management [12]. However, the willingness to participate in urban environmental activities was identified as an influential element in perceiving IGS favorably. To structure the integrative UGS planning for compact and green cities, a landscape ecological approach, governance processes, and public participation is required to adopt the demands of residents [54]. This suggests efforts to increase overall participatory green space management may lead to more positive perceptions of IGS in the future. Our findings corroborate previously proposed principles for participatory IGS management [12] and highlight the importance of non-IGS related experience in facilitating willingness to participate.

4.4. Limitations

This study has some limitations. Older residents (over 60) accounted for almost half of all respondents (44.7%). Therefore, it is assumed that the perception of the elderly has been reflected more strongly. However, this can be interpreted to provide a glimpse into the future Ichikawa is heading towards due to the rapid aging process ongoing in Japan. In an aging society, encouraging equality of outdoor activities and green life for the elderly is thus of increasing importance.

Another limitation was the number of 'undecided' responses to the perceived IGS in our survey responses. However, similar results by Rupprecht [12], despite using a different data collection method, suggest this may be typical for the study topic. While the reason could be a lack of interest in IGS, we find it more likely that the unfamiliarity of the concept makes expressing strong opinions difficult for residents. In the future, we propose testing a six-level Likert scale rather than a five-level Likert scale when surveying unfamiliar concepts, providing respondents with more nuanced ways to indicate weak agreement or disagreement while ensuring all respondents' opinions are reflected in the final results.

5. Conclusions

This study examined the potential of IGS as supplementary greenspace to meet the wellbeing needs of residents in the context of spatial and financial limitations in Ichikawa, Japan. Based on our findings, we conclude that IGS in Ichikawa is not disparate from green spaces that are recognized by residents, and has potential as a supplement for UGS. IGS can play a role in relieving the spatial and financial burden of governments and help them meet the needs of residents' comfortable lives. However, planners must consider ways to compensate for the fact that it may be difficult for residents with little UGS and related experience to perceive the potential of IGS. Therefore, when discussing IGS to resolve the inequality of green space provision, proposals should consider the perceptions of residents disadvantaged in terms of green space access to address this environmental justice issue. Another issue for planners to consider is the distinct spatial form of IGS. IGS is smaller than large-scale urban parks, and the continuity of space may be uncertain. IGS, however, is a result of spatially appearing by-products of human activities, scattered around the area where human activities take place. As our findings show, accessibility is one of IGS's most significant features and potential advantages—something planners can seek to leverage. This suggests that even though it may be difficult to provide users with the full functions of green space, such as an urban park, it can provide a minimum level of functions that can contribute towards meeting residents' needs in some parts of everyday life.

We conclude with some directions for future research based on our findings and limitations of our study in the hope they will contribute to furthering our understanding of IGS. Since close to half of our respondents were over 60 years old, we believe older adults' perception of IGS and its potential for them merits further investigation. Although our study was limited to Japan, represented by a shrinking and aging city, we suggest additional research in other Asian cities that share the issue of aging as an aspect affecting the quality of residents' lives, but which still experience rapid urban growth (e.g., Seoul). While a study in rapidly growing Brisbane, Australia suggested that IGS exists even when development pressure is high, IGS availability in Asian megacities is a topic that merits further study. Such a follow-up study of the availability of IGS should also consider recognition by older people in response to the increasingly aging Asian societies. Furthermore, in this context, IGS could be investigated as a relief not only for the elderly, but also people in lower socioeconomic groups who often experience unequal availability of green space; however, more data is needed on this topic. In addition, even though research on IGS has been increasing, support from the government and stakeholders is still limited because IGS's recreational use is contested by continuous development and land speculation pressure. Future work should thus investigate the direction of IGS's empirical development through perceptions towards IGS by residents and what role the government and urban planners play in how IGS are integrated into policies.

Author Contributions: M.K. conceived and designed the experiments, conducted the survey, analyzed the data, and wrote the initial draft; C.D.D.R. and K.F. provided support with research design and logistics; M.K. and C.D.D.R. revised and finalized the paper.

Funding: This research was supported by the FEAST Project (No. 14200116), Research Institute for Humanity and Nature (RIHN), and by JSPS KAKENHI Grant Numbers JP17K08179, JP17K15407.

Acknowledgments: We would like to thank all respondents for participating in this study.

Conflicts of Interest: The authors declare no conflict of interest. The founding sponsors had no role in the design of the study; in the collection, analyses, or interpretation of data; in the writing of the manuscript, and in the decision to publish the results.

References

1. United Nations. *World Urbanization Prospects: The 2018 Revision, Key Facts*; United Nations: New York, NY, USA, 2018.

2. WHO Regional Office for Europe. *Urban Green Spaces: A Brief for Action*; WHO Regional Office for Europe: Copenhagen, Denmark, 2017; pp. 1–24.

3. Hartig, T.; Evans, G.W.; Jamner, L.D.; Davis, D.S.; Gärling, T. Tracking restoration in natural and urban field settings. *J. Environ. Psychol.* **2003**, *23*, 109–123. [CrossRef]

4. Mackay, G.J.; Neill, J.T.; Richardson, E.A.; Mitchell, R.; Stigsdotter, U.K.; Palsdottir, A.M.; Burls, A.; Chermaz, A.; Ferrini, F.; Grahn, P.; et al. Does the outdoor environment matter for psychological restoration gained through running? *J. Environ. Psychol.* **2014**, *23*, 159–170. [CrossRef]

5. Matsuoka, R.H.; Kaplan, R. People needs in the urban landscape: Analysis of Landscape and Urban Planning contributions. *Landsc. Urban Plan.* **2008**, *84*, 7–19. [CrossRef]

6. Newton, J. Wellbeing and the Natural Environment: A Brief Overview of the Evidence. 2007, pp. 1–53. Available online: http://citeseerx.ist.psu.edu/viewdoc/download?doi=10.1.1.475.5690&rep=rep1&type=pdf (accessed on 28 August 2018).

7. Coley, R.L.; Sullivan, W.C.; Kuo, F.E. Where Does Community Grow? The Social Context Created by Nature in Urban Public Housing. *Environ. Behav.* **1997**, *29*, 468–494. [CrossRef]

8. Chiesura, A. The role of urban parks for the sustainable city. *Landsc. Urban Plan.* **2004**, *68*, 129–138. [CrossRef]

9. Jim, C.Y.; Chen, W.Y. Perception and attitude of residents toward urban green spaces in Guangzhou (China). *Environ. Manag.* **2006**, *38*, 338–349. [CrossRef] [PubMed]

10. Zhou, X.; Parves Rana, M. Social benefits of urban green space. *Manag. Environ. Qual.* **2012**, *23*, 173–189. [CrossRef]

11. Cowan, R.; Hill, D. *Creating Sustainable Urban Green Spaces*; Cabe Space: London, UK, 2005; ISBN 1846330009.

12. Rupprecht, C. Informal Urban Green Space: Residents' Perception, Use, and Management Preferences across Four Major Japanese Shrinking Cities. *Land* **2017**, *6*, 59. [CrossRef]

13. Yokohari, M.; Amati, M.; Bolthouse, J.; Kurita, H. Restoring urban fringe landscapes through urban agriculture:the Japanese experience. *Plan. Rev.* **2010**, *46*, 51–59. [CrossRef]

14. Ministry of Finance Japen. *Budget for Public Work Projects*; Ministry of Finance Japen: Tokyo, Japan, 2017.

15. Feltynowski, M.; Kronenberg, J.; Bergier, T.; Kabisch, N.; Łaszkiewicz, E.; Strohbach, M.W. Challenges of urban green space management in the face of using inadequate data. *Urban For. Urban Green.* **2018**, *31*, 56–66. [CrossRef]

16. Speer, J. Urban Interstices: The Aesthetics and the Politics of the In-Between. *Emot. Space Soc.* **2015**, *14*, 43–44. [CrossRef]

17. Doron, G.M. The Dead Zone and the Architecture of Transgression. *City* **2000**, *4*, 247–263. [CrossRef]

18. Shaw, P. The Qualities of Informal Space: (Re)appropriation within the informal, interstitial spaces of the city. In Proceedings of the Conference on Occupations: Negotiations with Constructed Space, Bright, UK, 2–4 July 2009; pp. 1–13.

19. Del Tredici, P. Spontaneous Urban Vegetation: Reflections of Change in a Globalized World. *Nat. Cult.* **2010**, *5*, 299–315. [CrossRef]

20. De Vries, S.; van Dillen, S.M.E.; Groenewegen, P.P.; Spreeuwenberg, P. Streetscape greenery and health: Stress, social cohesion and physical activity as mediators. *Soc. Sci. Med.* **2013**, *94*, 26–33. [CrossRef] [PubMed]

21. Anderson, E.C.; Minor, E.S. Vacant lots: An underexplored resource for ecological and social benefits in cities. *Urban For. Urban Green.* **2017**, *21*, 146–152. [CrossRef]
22. Németh, J.; Langhorst, J. Rethinking urban transformation: Temporary uses for vacant land. *Cities* **2014**, *40*, 143–150. [CrossRef]
23. Bonthoux, S.; Brun, M.; Di Pietro, F.; Greulich, S.; Bouché-Pillon, S. How can wastelands promote biodiversity in cities? A review. *Landsc. Urban Plan.* **2014**, *132*, 79–88. [CrossRef]
24. Jorgensen, A.; Tylecote, M. Ambivalent landscapes—Wilderness in the urban interstices. *Landsc. Res.* **2007**, *32*, 443–462. [CrossRef]
25. Rupprecht, C.D.D.; Byrne, J.A. Informal urban green-space: Comparison of quantity and characteristics in Brisbane, Australia and Sapporo, Japan. *PLoS ONE* **2014**, *9*, e99784. [CrossRef] [PubMed]
26. Pietrzyk-Kaszynska, A.; Czepkiewicz, M.; Kronenberg, J. Eliciting non-monetary values of formal and informal urban green spaces using public participation GIS. *Landsc. Urban Plan.* **2017**, *160*, 85–95. [CrossRef]
27. Botzat, A.; Fischer, L.K.; Kowarik, I. Unexploited opportunities in understanding liveable and biodiverse cities. A review on urban biodiversity perception and valuation. *Glob. Environ. Chang.* **2016**, *39*, 220–233. [CrossRef]
28. Rupprecht, C.D.D.; Byrne, J.A. Informal urban greenspace: A typology and trilingual systematic review of its role for urban residents and trends in the literature. *Urban For. Urban Green.* **2014**, *13*, 597–611. [CrossRef]
29. Unterweger, P.; Schrode, N.; Betz, O. Urban Nature: Perception and Acceptance of Alternative Green Space Management and the Change of Awareness after Provision of Environmental Information. A Chance for Biodiversity Protection. *Urban Sci.* **2017**, *1*, 24. [CrossRef]
30. Ichikawa City Urban Planning Division. *Urban Infra of Ichikawa Based on Data 2017*; Ichikawa City Urban Planning Division: Ichikawa, Japan, 2017.
31. Ichikawa City Urban Planning Division. *Ichikawa Urban Master Plan 2013*; Ichikawa City Urban Planning Division: Ichikawa, Japan, 2013.
32. Ichikawa City Urban Planning Division. *Ichikawa Green Master Plan 2004*; Ichikawa City Urban Planning Division: Ichikawa, Japan, 2004.
33. Trochim, W.; Donnelly, J.P. *Research Methods Knowledge Base*; Thomson Custom Publication: Mason, OH, USA, 2006.
34. Field, A.; Miles, J.; Field, Z. *Discovering Statistics Using SPSS*; Sage Publications Ltd: London, UK, 2013; Volume 81, ISBN 9781847879066.
35. George, D.; Mallery, P. *SPSS for Windows Step by Step A Simple Guide and Reference Answers to Selected Exercises*; Allyn & Bacon: Boston, MA, USA, 2003; p. 63. ISBN 9780205755615.
36. Clatworthy, J.; Hinds, J.; Camic, P.M. Gardening as a mental health intervention: A review. *Ment. Health Rev. J.* **2013**, *18*, 214–225. [CrossRef]
37. Newell, F.N.; Bülthoff, H.H. Categorical perception of familiar objects. *Cognition* **2002**, *85*, 113–143. [CrossRef]
38. Reyes, M.; Páez, A.; Morency, C. Walking accessibility to urban parks by children: A case study of Montreal. *Landsc. Urban Plan.* **2014**, *125*, 38–47. [CrossRef]
39. Ekkel, E.D.; de Vries, S. Nearby green space and human health: Evaluating accessibility metrics. *Landsc. Urban Plan.* **2017**, *157*, 214–220. [CrossRef]
40. Rojas, C.; Páez, A.; Barbosa, O.; Carrasco, J. Accessibility to urban green spaces in Chilean cities using adaptive thresholds. *J. Transp. Geogr.* **2016**, *57*, 227–240. [CrossRef]
41. Rupprecht, C.D.D.; Byrne, J.A.; Ueda, H.; Lo, A.Y. "It's real, not fake like a park": Residents' perception and use of informal urban green-space in Brisbane, Australia and Sapporo, Japan. *Landsc. Urban Plan.* **2015**, *143*, 205–218. [CrossRef]
42. De Vries, S.; Verheij, R.A.; Groenewegen, P.P.; Spreeuwenberg, P. Natural environments—Healthy environments? An exploratory analysis of the relationship between greenspace and health. *Environ. Plan. A* **2003**, *35*, 1717–1731. [CrossRef]
43. Maas, J. Green space, urbanity, and health: How strong is the relation? *J. Epidemiol. Community Health* **2006**, *60*, 587–592. [CrossRef] [PubMed]
44. Gong, F.; Zheng, Z.-C.; Ng, E. Modeling Elderly Accessibility to Urban Green Space in High Density Cities: A Case Study of Hong Kong. *Procedia Environ. Sci.* **2016**, *36*, 90–97. [CrossRef]
45. Sugiyama, T.; Thompson, C.W. Associations Between Neighborhood Open Space Attributes and Quality of Life for Older People in Britain. *Environ. Behav.* **2008**, *41*, 1–19. [CrossRef]

46. Takano, T. Urban residential environments and senior citizens' longevity in megacity areas: The importance of walkable green spaces. *J. Epidemiol. Community Health* **2002**, *56*, 913–918. [CrossRef] [PubMed]

47. WHO Regional Office for Europe. *Urban Green Spaces and Health: A Review of the Evidence*; World Health Organization: Geneva, Switzerland, 2016.

48. Harvey, J.A.; Chastin, S.F.M.; Skelton, D.A. Prevalence of sedentary behavior in older adults: A systematic review. *Int. J. Environ. Res. Public Health* **2013**, *10*, 6645–6661. [CrossRef] [PubMed]

49. Steptoe, A.; Shankar, A.; Demakakos, P.; Wardle, J. Social isolation, loneliness, and all-cause mortality in older men and women. *Proc. Natl. Acad. Sci. USA* **2013**, *110*, 5797–5801. [CrossRef] [PubMed]

50. Kweon, B.-S.; Sullivan, W.C.; Wiley, A.R. Green Common Spaces and the Social Integration of Inner-City Older Adults. *Environ. Behav.* **1988**, *30*, 832–858. [CrossRef]

51. Dennis, M.; James, P. User participation in urban green commons: Exploring the links between access, voluntarism, biodiversity and well being. *Urban For. Urban Green.* **2016**, *15*, 22–31. [CrossRef]

52. Patricia Hynes, H.; Howe, G. Urban horticulture in the contemporary united states: Personal and community benefits. *Acta Hortic.* **2004**, *643*, 171–181. [CrossRef]

53. Alaimo, K.; Packnett, E.; Miles, R.A.; Kruger, D.J. Fruit and Vegetable Intake among Urban Community Gardeners. *J. Nutr. Educ. Behav.* **2008**, *40*, 94–101. [CrossRef] [PubMed]

54. Artmann, M.; Bastian, O.; Grunewald, K. Using the concepts of green infrastructure and ecosystem services to specify leitbilder for compact and green cities-The example of the landscape plan of Dresden (Germany). *Sustainability* **2017**, *9*, 198. [CrossRef]

Article

Prioritizing Suitable Locations for Green Stormwater Infrastructure Based on Social Factors in Philadelphia

Zachary Christman [1,*], Mahbubur Meenar [1], Lynn Mandarano [2] and Kyle Hearing [2]

[1] Department of Geography, Planning, and Sustainability, School of Earth and Environment, Rowan University, Glassboro, NJ 08028, USA; meenar@rowan.edu
[2] Department of Planning and Community Development, Tyler School of Art, Temple University, Philadelphia, PA 19122, USA; lynn.mandarano@temple.edu (L.M.); kyle.hearing@temple.edu (K.H.)
* Correspondence: christmanz@rowan.edu; Tel.: +1-856-256-4810

Received: 30 October 2018; Accepted: 22 November 2018; Published: 26 November 2018

Abstract: Municipalities across the United States are prioritizing green stormwater infrastructure (GSI) projects due to their potential to concurrently optimize the social, economic, and environmental benefits of the "triple bottom line". While placement of these features is often based on biophysical variables regarding the natural and built environments, highly urbanized areas often exhibit either limited data or minimal variability in these characteristics. Using a case study of Philadelphia and building on previous work to prioritize GSI features in disadvantaged communities, this study addresses the dual concerns of the inequitable benefits of distribution and suitable site placement of GSI using a model to evaluate and integrate social variables to support decision making regarding GSI implementation. Results of this study indicate locations both suitable and optimal for the implementation of four types of GSI features: tree trenches, pervious pavement, rain gardens, and green roofs. Considerations of block-level site placement assets and liabilities are discussed, with recommendations for use of this analysis for future GSI programs.

Keywords: green stormwater infrastructure (GSI); social equity; site suitability modeling; geographic information systems; environmental justice; urban planning; Philadelphia

---

1. Introduction

The ability for vegetation to capture rainfall, mitigating overland sheet flows by promoting infiltration, reducing stormwater volume through transpiration, and filtering pollution through biological processes, has been well documented [1–3]. Simultaneously, managing rainfall poses unique challenges for many older urban areas in North America, which commonly utilize combined sewer systems (CSS) that manage both sewage and stormwater [4,5]. Intense rainfall events, either high volume or rapidly occurring, can exceed the capacity of these systems, leading to the diversion of untreated wastewater into rivers and other water bodies. Unsurprisingly, a growing number of cities across the United States have sought to leverage vegetation as a cost-effective means of mitigating the volume of stormwater and are increasingly allocating stormwater management funding to "green" as opposed to "grey" infrastructure projects [6,7].

Green stormwater infrastructure (GSI) refers to the suite of interventions, comprised of both natural and artificial materials, that utilize vegetation to slow or store surface water runoff, mitigating the volume rapidly reaching the CSS. The implementation of GSI depends largely on the intersection of properties associated with the physical and built environment as well as the priorities of municipal actors and community members [7,8]. This study specifically seeks to further develop the latter means of siting GSI through the development of a model for the balancing of social factors with the constraints of the built environment in prioritizing the implementation of GSI within Philadelphia, PA, USA. Specifically, answers to three major research questions were sought:

1.   What areas are suitable for GSI implementation, based on the physical constraints of the landscape and the goal of promoting equitable GSI distribution?
2.   How can social factors be used to prioritize and rank GSI site selection?
3.   How do virtual site observations corroborate modeled site suitability for GSI features?

In addressing the research questions outlined above, this study employs site suitability modeling based on an analytical hierarchical procedure of expert opinion on the influence of social factors to site GSI features. The study is built upon previous research done by Mandarano and Meenar [9] that identified Philadelphia census tracts for future GSI projects, prioritizing tracts with mid- to high-level of socio-economically disadvantaged residents but with high-level of community capacity. Results of this study identify locations across those high-priority census tracts for GSI features based on a variety of social factors. This study addresses gaps in existing procedures by utilizing social factors in urbanized regions that may have limited data or variability in characteristics of the natural and built environment commonly used to site green GSI.

2. Siting Green Stormwater Infrastructure

Municipalities across the United States are prioritizing GSI projects due to their "triple bottom line" benefits (e.g., social, economic, and environmental benefits). Fully realizing the social, economic, and environmental benefits of GSI necessitates a holistic methodology for siting GSI. The vast majority of research surrounding the implementation of GSI has largely ignored the social and economic benefits of GSI, instead choosing to focus principally on the environmental benefits. While this perspective for evaluating GSI improvements may effectively maximize pollutant reduction and runoff retention, it may not truly maximize the value of this infrastructure [10–14].

Adoption of an ecosystem services analysis of water supply investments, which seeks to comprehensively evaluate the environmental impact of infrastructure spending, reveals a grey area in the valuation of benefits, even before considering more abstract social benefits [15–18]. Attempts to measure and incorporate the social value of these ecosystem services underscore the difficulty in ascribing a singular value to any particular infrastructure, and challenges persist with respect to community education and adoption of any type of intervention [13,15,19]. In a pluralistic society such as the United States, engaging with and leveraging local institutions such as community groups and schools may represent a means of engaging with stakeholders over smaller-scale GSI projects [20].

Four major types of GSI interventions, varying in scale, structure, and operation are considered by this study: tree trenches, rain gardens, pervious pavements, and green roofs [7].

Tree Trenches are GSI elements comprised of a localized gap in the curbside impervious surface, planted with vegetation to temporarily store stormwater runoff from the street and sidewalk, enabling infiltration and evapotranspiration to decrease the volume entering the combined sewer system [6,21]. During major storm events, excess stormwater is directed to an existing stormwater inlet. Tree trench siting is contingent on the location of underground utilities and the required right-of-way, in addition to neighborhood physical and social factors discussed below. A tree trench may be as small as a single square meter along a sidewalk or an area of several dozen square meters along a building or pedestrian walkway.

Rain Gardens are extensive vegetated depressions that can collect water from surrounding impervious surfaces, thereby reducing velocity, promoting infiltration and evapotranspiration, and filtering pollutants from stormwater [7,22]. Connections to existing stormwater inlets manage excess volume during extreme precipitation events. Rain gardens necessitate an area of open space beyond the 100-year floodplain. Additionally, maintenance requirements of rain gardens make it highly desirable to site these features in proximity to institutions, such as schools or community centers, that may share in their upkeep. Spatial requirements for rain gardens vary from 10s to 1000s of square meters.

Pervious (or Permeable) Pavement interventions are a suite of design features that slow runoff water by promoting infiltration at the location that precipitation falls [7,23]. Though not required, pervious pavement is often accompanied by underground storage in urbanized areas, with connections to stormwater inlets for events when stormwater volumes exceed capacity. Additionally, the storage of water in these joint features must be cycled within days to avoid facilitating mosquitoes and other disease-harboring insects. The most suitable locations for the installation of pervious pavement are existing surface parking lots, outside the 100-year floodplain.

Green Roof features are additions to flat building roofs that reduce the velocity and volume of stormwater runoff by creating temporary storage and promoting evapotranspiration [7,24,25]. While newer green roof technology has reduced the structural burden of this type of infrastructure, thus expanding opportunities for retrofitting older structures, green roofs must feature downspouts connecting to the municipal wastewater system. Due to the financial and labor costs of installation and maintenance of these features, institutional capacity is also an important consideration.

Any GSI elements involving infiltration, including tree trenches, rain gardens, and pervious pavements, are affected by the underlying soil hydrology; however, in the densely built environment of many older cities, the longstanding disturbance of natural soil deposition limit the utility of this variable, and scant or unreliable data make this factor difficult to incorporate into models [26]. Similarly, topographic data can be used to site features, as tree trenches may benefit from a relatively flat topography to avoid inundation, while rain gardens and pervious pavements may be most effective when sited near slopes that accumulate a greater volume of stormwater [27]. In practice, the limited variation in slope of some cities, including Philadelphia, mean that this variable (i.e., slope) is similarly difficult to include in site selection models.

Concurrently, this built environment does exert a localized influence on the siting of GSI features. The presence of nearby impervious surfaces may support the effectiveness of GSI features promoting infiltration. In urbanized centers, impervious surface coverage may vary by block or parcel any change relatively rapidly necessitating the consideration of site-specific conditions [8,28–31]. In most cases, the engineering practices underlying GSI features necessitate proximity to a stormwater inlet to manage excess overflow and drainage [29]. Further, locating GSI features within an area served by a CSS more effectively furthers the ultimate goal of reducing stormwater infiltration and load on the system in order to reduce overflow events [5]. Finally, it is advisable to locate new GIS features beyond areas of expected inundations delineated by the US Federal Emergency Management Agency (FEMA) 100-year-flood map [32].

Beyond the physical requirements of any individual GSI intervention, many factors in the natural, built, and social environment must be considered in the selection of a suitable site for implementation [33–35]. Social factors that may be used to promote successful GSI interventions primarily involve metrics of proximity to institutions or community members who may directly or indirectly support them. Physical proximity to a partnered institution, such as a school, university, or recreation center may ease the financial and labor burdens of maintenance. Additionally, community organizations, such as Neighborhood Advisory Committees (NAC), can engage and educate residents, while also mustering volunteers for maintenance and promotion of these shared features. Further, these organizations may also provide long-term support through membership and leadership cycles. Characteristics of the specific target parcel and block ultimately carry critical importance that can be incorporated into a decision framework, both in the model and in the final site evaluation and validation [30]. Land tenure, including commercial or institutional ownership, areal requirements of the parcel, and local site configuration can enhance or diminish a site's suitability for GSI placement [29,36].

Though research has identified significant challenges associated with the incorporation of socioeconomic criteria in the siting of GSI, the need to consider these issues has been clearly demonstrated, albeit under the broader framework of sustainability [37]. At the same time, the social value of GSI has been empirically demonstrated; statistically significant decreases in burglaries and narcotics production and sale were found at multiple scales surrounding GSI in Philadelphia [8]. While crime statistics may be correctly identified as an insufficient proxy for capturing a broader range of socioeconomic factors, this research highlights the difficulty in quantifying these variables.

By constructing an analytical hierarchy framework for the prioritization of site-specific GSI [13], a model for maximizing the social and economic benefits of investments in GSI is proposed in this study. In the built environment, with limited variability or reliability of physical characteristics, social factors may exert a stronger influence on the successful implementation of GSI interventions, with regard to the impact of their placement to promote equitable access and impacts, especially among disadvantaged communities [38–40]. Previous work by Mandarano and Meenar [9] highlighted the inequitable distribution of GSI features across Philadelphia, due to their association especially with private, but also public, investment. That study integrated environmental justice and additional community context variables to identify high priority census tracts for new public-sector GSI implementations, based on community capacity.

3. Materials and Methods

This project was conducted in Philadelphia—the sixth most populous city of the United States—with an estimated population of 1.6 million residents [41] and a program to reduce the impervious surface due to the city's use of Combined Sewer Overflow (CSO) system [6,7,9]. The city is bordered to the east and south by the Delaware River and Bay and it is bisected by the Schuylkill River, with minimal topographical variability. Most of the city is covered by the CSS, and the neighborhoods that abut the rivers are within the FEMA 100-year flood zone. In Philadelphia, the Green City, Clean Waters program seeks to invest $2.4 billion over 25 years to capture 85% of the stormwater entering the sewer system [6,7]. Over the next 45 years, this program is expected to increase property values by $390 million by improving community quality of life and is estimated to prevent 140 fatalities by mitigating the impact of the urban heat island effect. Additionally, the program will employ 250 people, and improve air quality by absorbing an estimated 1.5 billion pounds of carbon dioxide annually [6,7]. This "triple bottom line" accounting of benefits was both essential to the program's adoption and unique amongst US municipalities [42].

Data for this analysis were obtained through several sources compiled via the OpenDataPhilly web portal [43]. High priority Census Tract data developed by Mandarano and Meenar [9] were thresholded to constrain the potential GSI implementation sites. Impervious surface and Combined Sewer Service Area data were prepared by the Philadelphia Water Department. Current tree plantings were obtained through the PhillyTreeMap, an implementation of OpenTreeMap built by Azavea with funding from the United States Department of Agriculture. The 100-year floodplain limits were prepared by the Federal Emergency Management Agency. Land cover data were produced by the City of Philadelphia with University of Vermont Spatial Analysis Laboratory. Building Footprint data were derived from Philadelphia License and Inspections database.

Prioritization and evaluation of suitable GSI implementation sites was conducted in four stages:

1) prior identification of high priority zones to achieve more equitable distribution based on community context and capacity, developed as part of previous research;
2) restriction of potential implementation sites based on constraints of the built and physical environment;
3) prioritization of potential implementation sites based on proximity to social criteria; and
4) virtual and in situ site evaluation for site feasibility and implementation considerations.

3.1. Prior Identification of High Priority Zones within Philadelphia

Data on high priority zones for future GSI locations in Philadelphia were collected from previous work by Mandarano and Meenar [9], who developed a strategy for public investment in GSI projects to achieve a more equitable distribution across the city. By following several methodological steps, they identified and prioritized US Census Tracts from the 2012 American Community Survey that included socio-economically disadvantaged populations but experienced high level of community capacity. First, using GIS-based raster overlay analysis, they identified and ranked disadvantaged census tracts by using community context variables that captured traditional environmental justice characteristics as well as other factors of disenfranchised communities, including demographic identity and rates of poverty, violent crime, vacant properties, single parent households, as well as a metric of income inequality. Next, they identified and ranked census tracts based on their level of community capacity. The community capacity variables incorporated measures of community capitals framework including educational attainment and median income of residents, presence of community organizations and number of residents who had participated in a civic engagement program, and the presence of public property and green space. Outputs from both analyses were overlaid to identify priority census tracts for equitable GSI distribution. Finally, the resulting prioritization scheme was ordinally ranked into five categories using a natural breaks classification method. In this study, the top two categories—indicating high levels of capacity and medium to high-levels of context/disadvantage—were used as a threshold to constrain all types of future GSI implementation. This process delimited priority zones for GSI implementation, which served to constrain this new study to further prioritize individual site selection for each type of GSI feature.

3.2. Restriction of Potential Implementation Sites

Locations within high priority GSI implementation zones were then further limited to CSS areas within the City of Philadelphia.

Tree Trenches: Potential GSI implementation zones for tree trenches were seeded with existing planting sites maintained by OpenTreeMap, under the rationale that any block that had any existing tree well or trench would satisfy requirements for sidewalk width, right-of-way, and be free of conflicting underground infrastructure, like water and electrical services.

Pervious Pavement: Potential GSI implementation zones for pervious pavement were seeded with existing surface parking lots, isolated from the map of impervious surfaces maintained by the Philadelphia Water Department, under the rationale that these features would be clear and accessible, with ongoing access following implementation. Additionally, sites were restricted to those beyond the limits of the FEMA 100-year floodplains along the Delaware and Schuylkill Rivers.

Rain Gardens: Potential GSI implementation zones for rain gardens were seeded with areas that were grass or bare earth, from the Urban Tree Canopy Assessment produced for Philadelphia by the University of Vermont Spatial Analysis Laboratory. These areas were further restricted to institutional land uses, under the rationale that these locations would have improved accessibility and maintenance. Based on anticipated size requirements of this type of implementation, areas smaller than 10 m^2 were removed, and zones were further restricted by being beyond the FEMA 100-year floodplains.

Green Roofs: Potential GSI implementation zones for green roofs were seeded with the City of Philadelphia Department of Licenses and Inspections building footprints database, which were further refined by those of civic and institutional ownership, isolated from the Philadelphia City Planning Commission, for more facile land use rights and access potential.

3.3. Prioritization of Features and Distances for Site Selection

The suitable sites identified above were prioritized based on their ranked proximity to a variety of features defined by the built and social environments using a site suitability analysis, as described below (Site suitability modeling is an analytical process to evaluate and integrate variables expressed in spatial data, in order to support decision making [44]. Variables regarding local characteristics or the proximity to a feature may be evaluated in comparison to a threshold value or a range of suitable values. Preparation of these variables is generally in two forms: constraints, which are binary criteria that impose strict inclusion or exclusion criteria, and factors, which can be evaluated to enable trade-off with other variables. Generally speaking, site suitability models incorporate constraints to identify the set of locations that may be considered for selection, and factors are used to prioritize selection within the set of all possible locations). The ranking of the influence of each factor on the ensuing selection was established through an analytical hierarchical weighting procedure based on the expert opinions of 16 professional planners, scholars, and municipal officials.

Factors chosen to act as indicators, signaling that a location may be a suitable location for GSI interventions were used through proximity metrics to the institutions or community members who may directly or indirectly support them, due to the potential to share in the financial and labor maintenance costs. Partner institutions, like schools, universities, or recreation centers, or community organizations, like NACs, may support the financial and labor costs of maintenance, educate and engage neighbors, and gather participants for events and education, especially through membership and leadership cycles. Linear distances to the nearest features were calculated in ArcMap 10.6 [45] using the NEAR function. Features for prioritized proximity included:

- Stormwater Inlets, necessary to drain excess water from all GSI features.
- Transit Stops, which act as community gateways. GSI and its accompanying vegetation have the potential to add vibrancy to the area and promote more than just environmentally sound stormwater management practices.
- Neighborhood Advisory Committees (NACs) are part of a Division of Housing and Community Development program whereby NACs lead and engage their neighborhoods in initiatives that align with the City's objectives including promoting sustainability, cultivating civic engagement, and ensuring residents have access to services. Nonprofits in eligible neighborhoods (low- and moderate-income) can partner with the City through this program; currently, 19 NACs operate throughout Philadelphia, and indicate a localized measure of neighborhood capacity.
- Schools, as GSI features offer potential educational value, and students exposed to GSI may learn about how it operates and the environmental implications of improperly managed stormwater runoff.
- Universities, as large institutions with financial resources and social obligations to pay for GSI on their campus or help maintain nearby GSI.
- Recreation Centers, which host civic events and programming, offer an opportunity to leverage the educational value of GSI and may support maintenance using existing staff and volunteers.

Opinions of participating experts was solicited to determine the functional distances from the features above to the GSI feature implementation site. Each feature has an optimal minimum distance and functionally limiting maximum distance. The range of each of these values was scaled linearly from 1 (best) to 0 (worst), with saturation points at each end of the scale beyond the minimum and maximum distances.

Based on the opinions of participating experts, the relative importance of the distances from each of the six criteria above were ranked from 1 (minimally influential) to 5 (critically influential), with 3 as a moderate or average influence. These were then scaled to a percentage of influence based on the sum of all factors per GSI feature.

The priority weights determined through this process were used in a linear combination to weigh the influence of all scaled distance factors for GSI site selection.

3.4. Virtual Site Observation and Validation

Sites for GSI implementation identified by the geospatial model outlined above were then inspected for potential feature installation using a checklist-based field assessment, which validated the efficacy of the model and identified relevant site-specific characteristics.

Visits were conducted virtually using the Street View service of Google Maps, which provides a ground-level 360° panoramic interface and orthogonal imagery from above to view roof and site configurations [46]. Locations extracted from the analysis in ArcMap were converted to a KML location document and integrated with the Google Maps platform for evaluation. At each location, factors commonly considered in the installation of GSI were evaluated by the virtual site assessment, including:

- slope of both the site and surrounding area (slopes in excess of 10 percent are generally excluded);
- impervious area of the parcels and roadways surrounding the site, to estimate the volume of potential drainage;

- spatial requirements for the potential GSI feature;
- proximity to existing GSI, in order to promote equitable distribution; and
- aesthetic considerations, including site visibility and potential obstructions.

Thirty-seven virtual site observations were conducted across the four types of proposed GSI interventions. For validation of site selection, all available regions per GSI type were merged (dissolved) into a single spatial feature, which was used to constrain the placement of the random points for site visits. A minimum distance of 161 m (528 feet, 1/10 mile, or approximately 1 major city block in Philadelphia) between points was established to avoid issues of proximity and spatial autocorrelation. This process ensured 10 independent validation points for tree trenches, pervious pavement, and green roofs. For rain gardens, the minimum distance was halved, but only 7 spatially independent points were possible due to the clustering of high priority locations. Thus, a total of 37 points comprised the extent of virtual site observation and validation.

4. Results

4.1. Prior Identification of High Priority Zones within the City of Philadelphia

Based on the selection of the two highest ordinally ranked sets of census tracts following the analysis by Mandarano and Meenar [9], approximately 13.00 km^2 of the city was isolated for further site selection and prioritization. These areas are shown in shaded grey in Figure 1.

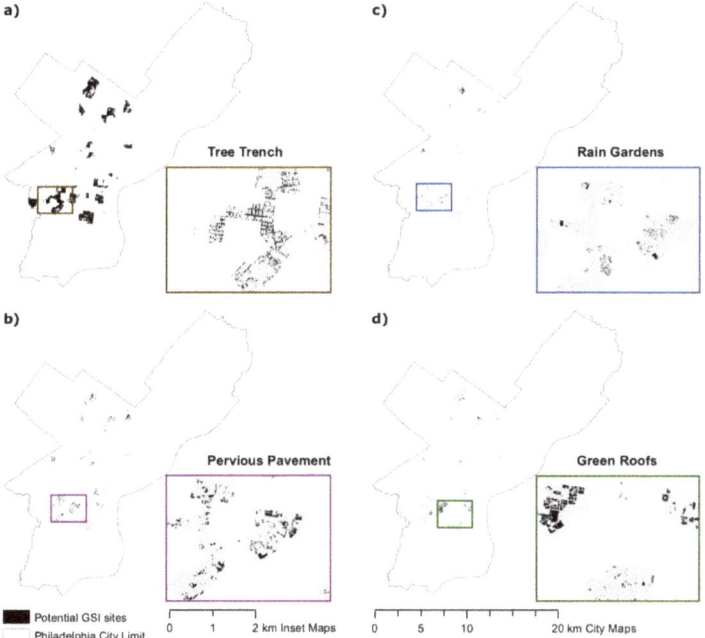

Figure 1. The distribution of potential GSI implementation sites by type of feature, across the City of Philadelphia.

4.2. Restriction of Potential Implementation Sites

Based on the constraints applied to the implementation of GSI features, a suite of suitable locations were identified. A set of 7850 locations of existing street trees were identified, representing locations for which there was adequate sidewalk width without interference of underground utility infrastructure.

Individual street tree wells may be linked for tree trenches or that areas along the same block may be selected for tree trench installation. Potential pervious pavement implementation sites included 527 parcels identified as suitable surface parking lots, totaling 0.554 square kilometers. Areas of grass or bare earth suitable for the installation of rain gardens yielded 1270 potential implementation sites, totaling 0.225 square kilometers. Building footprints with civic and institutional ownership included 833 potential implementation sites for green roofs, totaling 0.324 square kilometers. These areas are the black features for each type in Figure 1.

4.3. Prioritization of Features and Distances for Site Selection

Priority weights based on expert opinion are listed in Table 1. These weights were used in the linear combination to rank the suitable GSI implementation sites, which are depicted in Figure 2.

Table 1. Weighting of proximity to socially relevant features and their functional distances.

Feature	Tree Trench	Pervious Pavement	Rain Garden	Green Roof	Distance Range, Min-Max (m [feet])
Sewer Inlet	12.00%	41.70%	17.20%	23.50%	0–4.57 [0–15]
Transit Stop	16.00%	8.30%	13.80%	5.90%	6.10–15.24 [20–50]
School	20.00%	8.30%	17.20%	23.50%	15.24–762 [50–2500]
N.A.C.	16.00%	8.30%	17.20%	5.90%	0–457.2 [0–1500]
University	20.00%	25.00%	17.20%	23.50%	15.24–4828.0 [50–15840, 3 miles]
Rec. Center	16.00%	8.30%	17.20%	17.60%	15.24–304.8 [50–1000]

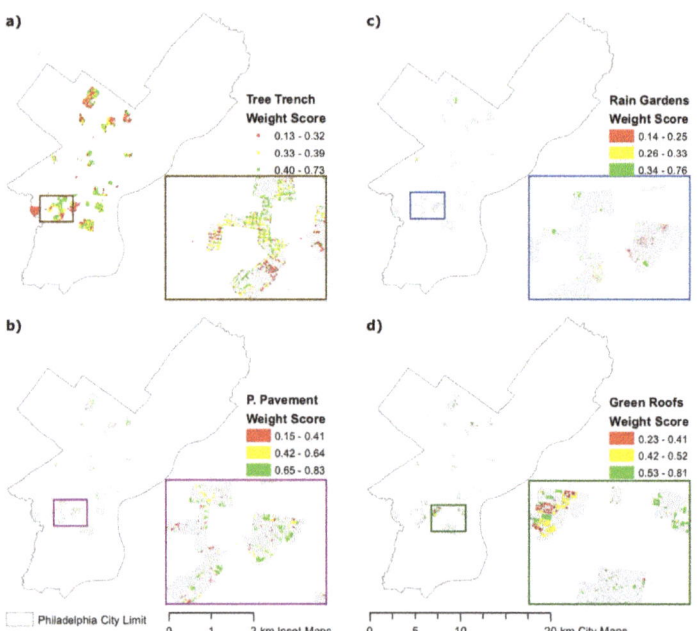

Figure 2. Ranked prioritization of potential GSI implementation sites by type of feature, across the City of Philadelphia.

Based on the weighted linear combination of site prioritization characteristics, a suitability score was calculated for each potential location for GSI implementation. Site scores ranged from 0 (least suitable) to 1 (most suitable). Results were categorized into three tiers by quantile classification, with full results in Figure 2. Scores for sites with the highest suitability ranged from 0.40–0.73 for tree trenches, 0.65–0.83 for pervious pavement, 0.53–0.81 for green roofs, and 0.34–0.76 for rain gardens.

4.4. Virtual Site Observation and Validation

Of the 37 virtual site observations, 25 (67%) of locations had the available 360° imagery dated in 2017, with the remainder in 2016 (5 locations), 2014 (6 locations), and 2009 (1 location). All imagery were accessed in August, 2018. Figure 3 depicts a potential GSI implementation site imagery and the distribution of sites across the high-priority Census Tracts of the study area.

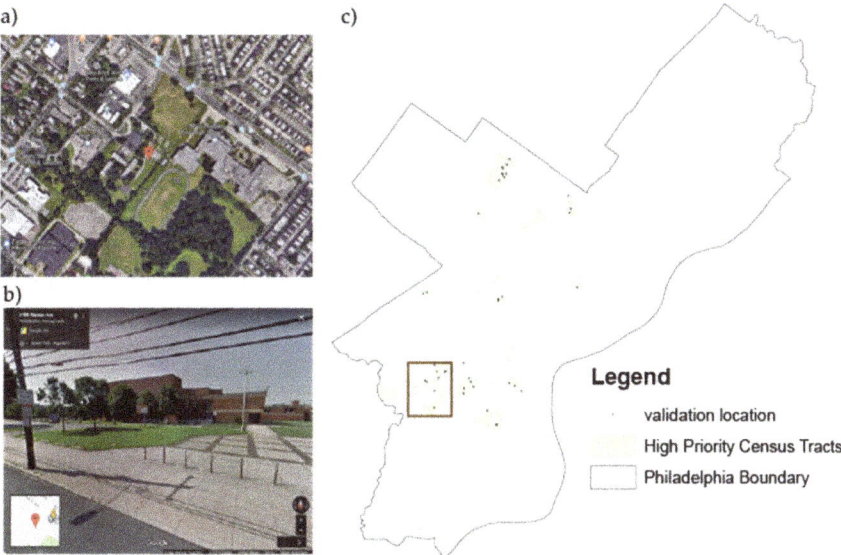

Figure 3. (**a**) sample location for potential siting of GSI feature in Google orthographic imagery; (**b**) StreetView 360° imagery; (**c**) location of all validation points, within previously established high priority census tracts.

At each location, a survey of the types and conditions of properties and streetside infrastructure was conducted through a visual examination of the orthographic and streetside 360° imagery, tallied in a standard survey questionnaire (Appendix B). Residential properties were visible at 62% of locations, with 96% described as moderately- or very- well-kept, on a 4-point Likert scale. Commercial properties were visible at 46% of locations, with 94% of those at least moderately well-kept. Industrial properties were visible at only 11% of locations, with 100% at least moderately well-kept. Other buildings were present at 78% of locations, with 97% at least moderately well-kept.

Regarding the streetside infrastructure: yards (vegetation or other aesthetic implementations) and sidewalks were noted at all sites, with 89% and 81%, respectively, at least moderately well-kept, free of litter, and not in obvious disrepair. Recreation facilities were visible at 22% of locations and parking lots were visible at 49% of locations; all of these features were at least moderately well-kept.

Of the sites virtually observed, only one (3%) had a slope that exceeded 2% but less than 5%; for all other locations, slopes were 0–2%. Sites varied widely regarding the presence of impervious surfaces, which were estimated for the streetside structure, adjacent lots, and institutional lots, and results are shown below in Table 2.

Table 2. Percent impervious surface surrounding high-priority GSI implementation sites.

Zone	Percent Impervious				
	0–25%	25–50%	50–75%	75–100%	Cannot Determine
Streetside structure	0.00%	21.62%	27.03%	51.35%	0.00%
Adjacent lots	5.41%	18.92%	27.03%	48.65%	0.00%
Institutional lots	5.41%	24.32%	8.11%	32.43%	29.73%

In addition to the type of GSI feature indicated by the suitability model, each site was also evaluated for its suitability for all GSI feature types. The suitability for the indicated GSI feature type at each location had high correspondence: 100% of the tree trench and pervious pavement sites, 90% of the green roof sites, and 85.71% of the rain garden sites were deemed suitable for their respective feature implementations. Overall, it was estimated that 95% of all sites visited for any potential GSI type had sufficient space in the nearby area for a tree trench GSI feature. For the placement of pervious pavement, 51% had sufficient space, and for the placement of a rain garden, 35% had sufficient space. Green roof placement was assessed by the presence of a visible flat roof in an adjacent commercial or institutional property, and 38% had sufficient space for this GSI feature.

Finally, addressing the question of whether the anticipated GSI use was found to be suitable for the observed location. In the virtual observation site examination, the potential suitability for each of the four GSI types was evaluated at every location, to account for potential local site substitution based on community and design implementation factors. Full results are listed in Table 3.

Table 3. Alignment of anticipated use with observed suitability from virtual field observation.

Type of GSI	Observed Suitability (Virtual Field Observation/Validation)			
	Tree Trench	Pervious Pavement	Green Roof	Rain Garden
Tree trench	100.00%	90.00%	90.00%	100.00%
Pervious pavement	20.00%	100.00%	40.00%	42.86%
Green roof	10.00%	30.00%	90.00%	14.29%
Rain garden	30.00%	20.00%	20.00%	85.71%

5. Discussion

Though the social and economic benefits associated with GSI have been well documented [8,37], efforts to maximize these benefits have been limited. Building on research identifying census tracts in Philadelphia with indicators for disadvantage as well as a high community capacity for collective action [9], this prioritization framework offers a means of maximizing the triple bottom line benefits of GSI while considering the equity and long-term viability of these investments.

While the focus of municipal agencies may be, understandably, on the environmental impact of the proposed GSI, these analyses may benefit from the consideration of a broader range of criteria. Potential advantages include: maximizing the social and economic benefits of GSI; equitably distributing GSI; distinguishing amongst similar candidate parcels (common in urban areas); and leveraging community capacity to facilitate the ongoing maintenance of GSI. As municipalities struggle to meet federal requirements for reducing CSOs, resources are often divided across "green" and "grey" infrastructure projects [42]. When GSI projects are sited to maximize social and economic benefits, they may be viewed in more economically favorable terms, which may, in turn, accelerate adoption of sustainable municipal stormwater management practices.

This study represents a starting point for parties interested in the social and economic benefits of GSI. Previous work demonstrated the practicality of prioritizing social equity and community capacity while raising important questions for future research to answer [9]. The framework and subsequent prioritization developed through this research have immediate relevance for those working to install GSI throughout Philadelphia in addition to long-term implications for municipalities and their agencies seeking to maximize the impact of GSI projects. Further research must also address the benefits and perception thereof by residents and local officials, which can influence the perceived success and future public support for GSI projects.

Within Philadelphia, validation and local site examination suggests that the process of restricting potential sites based on the distance-based social features of the site was an effective means of identifying sites for prioritization. Though it is difficult to even anticipate the effectiveness of the subsequent priority ranking process, the variability of site rankings resulting from the analysis of expert opinion suggests that this process effectively distinguished among comparable candidate sites.

Results of this study are significant to the fields of environmental planning and management, restorative sustainability, and environmental justice, by enhancing methods to identify sites for GSI implementation in highly urbanized areas and disadvantaged communities. First and foremost, the findings demonstrate the utility of a site suitability model that employs social factors coupled with virtual field observations to evaluate siting of GSI projects in Philadelphia. The site suitability model developed is unique in its approach to identify GSI sites in highly urbanized areas using physical constraints of the built environment coupled with social factors. A lack of variability or reliable data limit the utility of environmental data for siting GSI in urban areas that may have limited undisturbed natural landscapes. In response, this model used built environmental characteristics, including locations of trees and vacant tree wells, parking lots, grassy areas, and building footprints for the respective GSI features to be implemented. While proximity to existing stormwater infrastructure is a conventional criterion used in siting GSI, this study also used expert opinion to identify and rank other proximity criteria. While the expert opinions were calibrated by the conditions in Philadelphia, the methodology of incorporating such social factors is transferrable.

The approach to validating the sites generated from the suitability model also facilitates the site selection process. The use of virtual site observations of the suitability model results demonstrated that the model generated locations that ranged from 85.71% to 100% suitable for the proposed GSI project. In addition, it is important to note the sites suitable for tree trenches also were appropriate locations for other types of GSI projects. The novel use of a virtual approach to field validation offers the potential for substantial savings in in time and resources.

This study also contributes to strategies to bring nature-based solutions to environmental justice communities through a two-tiered method to identified locations for GSI projects with the potential to deliver triple bottom line benefits. Building on previous work that developed a strategy to prioritize Census tracts for equitable distribution of GSI in Philadelphia [9], this study employed a suitability model with proximity criteria and weights based on expert opinion. Together, these methods identified block-level suitable locations with community capacity bolstered by transit stops, NACs, schools, and recreation centers.

This study is not without limitations. While the principles upon which this site-specific prioritization framework is built may be broadly transferrable, the physical and social characteristics meriting consideration may vary significantly. With regards to physical characteristics, slope and hydrologic soil grouping, variables excluded from analysis in the case of Philadelphia, could prove critically important if this methodology was to be utilized elsewhere. Similarly, NACs are unique to Philadelphia, and Universities may not provide a comparable institutional presence in another region of analysis. Finally, as the form and availability of data influenced the design of this prioritization framework, it would be expected to influence subsequent applications of the framework across geographies in different ways, resulting in the consideration of new features of the social and built environments. It is also important to note that while this framework includes expert opinion to inform the site selection process at the local level, it does not include direct input from the impacted community. Ultimately, community preference for and knowledge of the appropriate location and type of GSIs should be incorporated to further guide implementation and this can be addressed in a future study.

Longitudinal research, comparing the actual benefits of GSI elements with their prioritization score would be required to measure the effectiveness of this particular prioritization framework; however, the complexity of measuring these variables and the temporal scope required to conduct such an analysis will likely prove prohibitive. Nevertheless, demonstrating the replicability of this methodology, appropriately adjusted for a new geography, is warranted as a means of validating this approach to the prioritization of GSI, incorporating features of both the social and built environment. Such replication would further this model of accounting for the positive externalities of GSI and serve as a means of furthering its propagation. Further research should incorporate the perceptions of GSI features by the heterogeneous communities across the urban context and how these features may promote or compete with community well-being, heritage, and future development.

6. Conclusions

This study demonstrated a framework for siting GSI features in the context of an urbanized area with limited variability in the characteristics of the natural and built environments commonly used. Through this analysis, locations for four different types of GSI features were identified within the high-priority census tracts identified by previous work, which balanced environmental justice and community capacity factors [9]. This study incorporated the opinions of experts to prioritize and delimit distances of influence for GSI siting, which were validated and further contextualized using virtual field observations, with potentially transferrable insights for other urbanized areas that seek to control stormwater impacts on aging infrastructure.

Author Contributions: Conceptualization, L.M., M.M., and Z.C.; methodology, M.M., K.H., and Z.C.; data curation, K.H.; analysis, K.H. and Z.C.; validation, Z.C.; writing—original draft preparation, M.M., Z.C., K.H., and L.M.; writing—review and editing, Z.C., M.M., K.H., and L.M.; visualization, Z.C. and K.H.

Funding: This research has been supported by a grant from the U.S. Environmental Protection Agency's Science to Achieve Results (STAR) program under Assistance Agreement No._835557. It has not been formally reviewed by EPA. The views expressed in this document are solely those of authors and do not necessarily reflect those of the Agency. EPA does not endorse any products or commercial services mentioned in this publication.

Acknowledgments: We would like to acknowledge Rowan University students Cassie Shugart, Robert Kearney, and Donald Morrison for their support in the beginning stage of this project. We thank Philadelphia area GSI experts for their valuable feedback on the suitability model criteria selection process.

Conflicts of Interest: The authors declare no conflict of interest.

Appendix A. Questions asked of Experts regarding Influence and Relative Importance of Factors Facilitating the Prioritization of Green Stormwater Infrastructure

The following questions were asked in identical form for each of the four types of GSI features (tree trenches, rain gardens, pervious pavement, and rain gardens):

1. Regarding the distance to the nearest stormwater inlet, what is the minimum optimal distance (in feet), and is there a range of minimum values that would all be optimal (e.g., a stormwater inlet closer than 20 feet is ideal and equivalent in terms of cost & efficacy)? (text input)
2. Now, regarding the distance to the nearest stormwater inlet, what is the maximum optimal distance (in feet) beyond which this factor no longer influences the choice or decision framework for this GSI element (e.g., a stormwater inlet more than 100 feet away is irrelevant)? (text input)
3. Regarding the distance to the nearest transit stop, what is the minimum optimal distance (in feet), and is there a range of minimum values that would all be optimal (e.g., a transit stop Proximity of nearest recreation center
4. Now, regarding the distance to the nearest transit stop, what is the maximum optimal distance (in feet) beyond which this factor no longer influences the choice or decision framework for this GSI element (e.g., a transit stop more than 1,000 feet away is irrelevant)?
5. Regarding the distance to the nearest Neighborhood Advisory Committee, what is the minimum optimal distance (in feet), and is there a range of minimum values that would all be optimal?

6. Now, regarding the distance to the nearest Neighborhood Advisory Committee, what is the maximum optimal distance (in feet) beyond which this factor no longer influences the choice or decision framework for this GSI element?

7. Regarding the distance to the nearest school, what is the minimum optimal distance (in feet), and is there a range of minimum values that would all be optimal?

8. Now, regarding the distance to the nearest school, what is the maximum optimal distance (in feet) beyond which this factor no longer influences the choice or decision framework for this GSI element?

9. Regarding the distance to the nearest university, what is the minimum optimal distance (in feet), and is there a range of minimum values that would all be optimal?

10. Now, regarding the distance to the nearest university, what is the maximum optimal distance (in feet) beyond which this factor no longer influences the choice or decision framework for this GSI element?

11. Regarding the distance to the nearest recreation center, what is the minimum optimal distance (in feet), and is there a range of minimum values that would all be optimal?

12. Now, regarding the distance to the nearest recreation center, what is the maximum optimal distance (in feet) beyond which this factor no longer influences the choice or decision framework for this GSI element?

13. In siting a tree trench, how would you rank the relative importance of these factors? (choice matrix

    a.   Factors

        i.   Proximity of nearest stormwater inlet
        ii.   Proximity of nearest transit stop
        iii.   Proximity of nearest Neighborhood Advisory Committee
        iv.   Proximity of nearest school
        v.   Proximity of nearest university
        vi.   Proximity of nearest recreation center

    b.   Influence

        i.   Not at all important
        ii.   Some importance
        iii.   Somewhat important
        iv.   Very important
        v.   Most important

Appendix B. Green Stormwater Infrastructure site validation and micro-characteristics survey

1. Number of feature (from stratified random sample scheme) (text input)
2. Latitude/Longitude of observed point (text input)
3. Street address of observed point (text input)
4. Date of Google Street View imagery (Month/year)
5. Type of Site under consideration at this location (choice matrix)

    a.   Tree trench
    b.   Pervious pavement
    c.   Rain garden
    d.   Green roof

6. Slope limitations

    a.   0–2%

    b.   2–5%

    c.   5–10%

    d.   10+%

7.   Current impervious surface (choice matrix)

    a.   Types

        i.   Street-side structure

        ii.   Adjacent lots

        iii.   Institutional lots

    b.   Ranges

        i.   0–25%

        ii.   25–50%

        iii.   50–75%

        iv.   75–100%

        v.   cannot determine

8.   Assets of site (checkbox)

    a.   Site has good visibility from street

    b.   Institution is visible at this location

9.   Asset comments (text input)

10.   Liabilities of site (checkbox)

    a.   Visible trash

    b.   Visible disrepair of adjacent properties

11.   Liability comments (text input)

12.   Block characteristics (choice matrix))

    a.   Types

        i.   Residential properties

        ii.   Commercial properties

        iii.   Industrial properties

        iv.   Other buildings

        v.   Yards and surrounding areas

        vi.   Sidewalks

        vii.   Recreational facilities

        viii.   Parking lots

        ix.   Vacant lots or unused marked space

    b.   Conditions

        i.   Not present

        ii.   Poor/badly deteriorated (extensive damage, neglect)

        iii.   Fair condition (e.g., peeing paint; needs repair)

        iv.   Moderately well-kept

        v.   Very well-kept (no visible problems)

        vi.   Cannot determine from view

13.   Is there sufficient space at this site for (choice matrix)

    a.   Types

           i.   Tree trench

           ii.   Pervious pavement

           iii.   Green roof

           iv.   Rain garden

    b.   Responses

           i.   Yes

           ii.   No

           iii.   Unclear from observation

14. Comments (text input)

References

1. Muerdter, C.P.; Wong, C.K.; LeFevre, G.H. Emerging investigator series: The role of vegetation in bioretention for stormwater treatment in the built environment: Pollutant removal, hydrologic function, and ancillary benefits. *Environ. Sci.-WATER Res. Technol.* **2018**, *4*, 592–612. [CrossRef]
2. Pennino, M.J.; McDonald, R.I.; Jaffe, P.R. Watershed-scale impacts of stormwater green infrastructure on hydrology, nutrient fluxes, and combined sewer overflows in the mid-Atlantic region. *Sci. Total Environ.* **2016**, *565*, 1044–1053. [CrossRef] [PubMed]
3. Li, C.; Fletcher, T.D.; Duncan, H.P.; Burns, M.J. Can stormwater control measures restore altered urban flow regimes at the catchment scale? *J. Hydrol.* **2017**, *549*, 631–653. [CrossRef]
4. Brombach, H.; Weiss, G.; Fuchs, S. A new database on urban runoff pollution: Comparison of separate and combined sewer systems. *Water Sci. Technol.* **2005**, *51*, 119–128. [CrossRef] [PubMed]
5. United States Environmental Protection Agency Office of Water. *Report to Congress: Impacts and Control of CSOs and SSOs*; EPA B33-R-04-001; Office of Water: Washington, DC, USA, August 2004.
6. LaDuca, A.; Kosko, J. *Getting to Green: Paying for Green Infrastructure–Financing Options and Resources for Local Decision-Makers*; EPA 842-R-14-005; United States Environmental Protection Agency: Washington, DC, USA, 2014.
7. Philadelphia Water Department. *Green Infrastructure Maintenance Manual*; Green City Clean Waters; Version 2.0; Philadelphia Water Department: Philadelphia, PA, USA, September 2016.
8. Kondo, M.C.; Low, S.C.; Henning, J.; Branas, C.C. The Impact of Green Stormwater Infrastructure Installation on Surrounding Health and Safety. *Am. J. Public Health* **2015**, *105*, e114–e121. [CrossRef] [PubMed]
9. Mandarano, L.; Meenar, M. Equitable distribution of green stormwater infrastructure: A capacity-based framework for implementation in disadvantaged communities. *Local Environ.* **2017**, *22*, 1338–1357. [CrossRef]
10. Tsihrintzis, V.A.; Hamid, R. Modeling and Management of Urban Stormwater Runoff Quality: A Review. *Water Resour. Manag.* **1997**, *11*, 136–164. [CrossRef]
11. Collins, K.A.; Lawrence, T.J.; Stander, E.K.; Jontos, R.J.; Kaushal, S.S.; Newcomer, T.A.; Grimm, N.B.; Cole Ekberg, M.L. Opportunities and challenges for managing nitrogen in urban stormwater: A review and synthesis. *Manag. Denitrification Hum. Domin. Landsc.* **2010**, *36*, 1507–1519. [CrossRef]
12. Shafique, M.; Kim, R.; Kyung-Ho, K. Green Roof for Stormwater Management in a Highly Urbanized Area: The Case of Seoul, Korea. *Sustainability* **2018**, *10*. [CrossRef]
13. Young Kevin, D.; Kibler David, F.; Benham Brian, L.; Loganathan, G.V. Application of the Analytical Hierarchical Process for Improved Selection of Storm-Water BMPs. *J. Water Resour. Plan. Manag.* **2009**, *135*, 264–275. [CrossRef]
14. Wang, R.; Eckelman, M.J.; Zimmerman, J.B. Consequential Environmental and Economic Life Cycle Assessment of Green and Gray Stormwater Infrastructures for Combined Sewer Systems. *Environ. Sci. Technol.* **2013**, *47*, 11189–11198. [CrossRef] [PubMed]
15. Kandulu, J.M.; Connor, J.D.; MacDonald, D.H. Ecosystem services in urban water investment. *J. Environ. Manage.* **2014**, *145*, 43–53. [CrossRef] [PubMed]
16. Dagenais, D.; Thomas, I.; Paquette, S. Siting green stormwater infrastructure in a neighbourhood to maximise secondary benefits: Lessons learned from a pilot project. *Landsc. Res.* **2017**, *42*, 195–210. [CrossRef]

17. Cox, J. "Green Stormwater Infrastructure"; *Parks & Recreation*. February 2018. Available online: https://www.nrpa.org/parks-recreation-magazine/2018/february/green-stormwater-infrastructure/ (accessed on 25 November 2018).

18. Cai, G.; Du, M.; Xue, Y. Monitoring of urban heat island effect in Beijing combining ASTER and TM data. *Int. J. Remote Sens.* **2011**, *32*, 1213–1232. [CrossRef]

19. Kati, V. Hotspots, complementarity or representativeness? designing optimal small-scale reserves for biodiversity conservation. *Biol. Conserv.* **2004**, *120*, 471–480. [CrossRef]

20. Kati, V.; Jari, N. Bottom-up thinking—Identifying socio-cultural values of ecosystem services in local blue–green infrastructure planning in Helsinki, Finland. *Land Use Policy* **2016**, *50*, 537–547. [CrossRef]

21. Jayasooriya, V.M.; Ng, A.W.M. Tools for Modeling of Stormwater Management and Economics of Green Infrastructure Practices: A Review. *Water. Air. Soil Pollut.* **2014**, *225*, 2055. [CrossRef]

22. Elliott, A.H.; Trowsdale, S.A. A review of models for low impact urban stormwater drainage. *Spec. Sect. Adv. Technol. Environ. Model.* **2007**, *22*, 394–405. [CrossRef]

23. Cheng, M.-S.; Zhen, J.X.; Shoemaker, L. BMP decision support system for evaluating stormwater management alternatives. *Front. Environ. Sci. Eng. China* **2009**, *3*, 453. [CrossRef]

24. Stovin, V. The potential of green roofs to manage Urban Stormwater. *Water Environ. J.* **2010**, *24*, 192–199. [CrossRef]

25. VanWoert, N.D.; Rowe, D.B.; Andresen, J.A.; Rugh, C.L.; Fernandez, R.T.; Xiao, L. Green Roof Stormwater Retention This paper is a portion of a thesis submitted by N.D. VanWoert. *J. Environ. Qual.* **2005**, *34*, 1036–1044. [CrossRef] [PubMed]

26. U.S. Department of Agriculture; Natural Resources Conservation Service. *Web Soil Survey*. Available online: https://websoilsurvey.sc.egov.usda.gov/App/HomePage.htm (accessed on 25 November 2018).

27. Pennsylvania Department of Environmental Protection Bureau of Watershed Management; *Pennsylvania Stormwater Best Management Practices Manual*; 363-0300-002. Available online: http://pecpa.org/wp-content/uploads/Stormwater-BMP-Manual.pdf (accessed on 25 November 2018).

28. Pataki, D.E.; Carreiro, M.M.; Cherrier, J.; Grulke, N.E.; Jennings, V.; Pincetl, S.; Pouyat, R.V.; Whitlow, T.H.; Zipperer, W.C. Coupling biogeochemical cycles in urban environments: Ecosystem services, green solutions, and misconceptions. *Front. Ecol. Environ.* **2011**, *9*, 27–36. [CrossRef]

29. Porse, E. Open data and stormwater systems in Los Angeles: Applications for equitable green infrastructure. *Local Environ.* **2018**, *23*, 505–517. [CrossRef]

30. Eaton, T.T. Approach and case-study of green infrastructure screening analysis for urban stormwater control. *J. Environ. Manag.* **2018**, *209*, 495–504. [CrossRef] [PubMed]

31. Liu, W.; Chen, W.; Peng, C. Influences of setting sizes and combination of green infrastructures on community's stormwater runoff reduction. *Ecol. Manag. Hum.-Domin. Urban Reg. Ecosyst.* **2015**, *318*, 236–244. [CrossRef]

32. Federal Emergency Management Agency *Flood Zones*. Available online: https://www.fema.gov/flood-zones (accessed on 25 November 2018).

33. Green, O.O.; Shuster, W.D.; Rhea, L.K.; Garmestani, A.S.; Thurston, H.W. Identification and Induction of Human, Social, and Cultural Capitals through an Experimental Approach to Stormwater Management. *Sustainability* **2012**, *4*. [CrossRef]

34. Boyle, L.; Michell, K.; Viruly, F. A Critique of the Application of Neighborhood Sustainability Assessment Tools in Urban Regeneration. *Sustainability* **2018**, *10*. [CrossRef]

35. Keeley, M.; Koburger, A.; Dolowitz, D.P.; Medearis, D.; Nickel, D.; Shuster, W. Perspectives on the Use of Green Infrastructure for Stormwater Management in Cleveland and Milwaukee. *Environ. Manag.* **2013**, *51*, 1093–1108. [CrossRef] [PubMed]

36. Garcia-Cuerva, L.; Berglund, E.Z.; Rivers, L., III. An integrated approach to place Green Infrastructure strategies in marginalized communities and evaluate stormwater mitigation. *J. Hydrol.* **2018**, *559*, 648–660. [CrossRef]

37. Sahely, H.R.; Kennedy, C.A.; Adams, B.J. Developing sustainability criteria for urban infrastructure systems. *Can. J. Civ. Eng.* **2005**, *32*, 72–85. [CrossRef]

38. Schilling, J.; Logan, J. Greening the Rust Belt: A Green Infrastructure Model for Right Sizing America's Shrinking Cities. *J. Am. Plann. Assoc.* **2008**, *74*, 451–466. [CrossRef]

39. Wolch, J.R.; Byrne, J.; Newell, J.P. Urban green space, public health, and environmental justice: The challenge of making cities 'just green enough'. *Landsc. Urban Plan.* **2014**, *125*, 234–244. [CrossRef]

40. Mell, I.C. Can green infrastructure promote urban sustainability? *Proc. Inst. Civ. Eng.-Eng. Sustain.* **2009**, *162*, 23–34. [CrossRef]
41. United States Census Bureau. *American Community Survey 1-Year Estiamtes*; American Community Sruvey; United States Census Bureau: Suitland, MD, USA, 2017.
42. Hopkins, K.G.; Grimm, N.B.; York, A.M. Influence of governance structure on green stormwater infrastructure investment. *Environ. Sci. Policy* **2018**, *84*, 124–133. [CrossRef]
43. *OpenDataPhilly* web portal. Available online: https://www.opendataphilly.org (accessed on 25 November 2018).
44. Eastman, J.R.; Jiang, H.; Toledano, J. Multi-criteria and multi-objective decision making for land allocation using GIS. In *Multicriteria Analysis for Land-Use Management*; Beinat, E., Nijkamp, P., Eds.; Springer: Dordrecht, The Netherlands, 1998; pp. 227–251. ISBN 978-94-015-9058-7.
45. ESRI 2018 *ArcGIS Desktop Release 10.6*; Redlands, CA: Environmental Systems Research Institute. Available online: http://desktop.arcgis.com/en/arcmap/ (accessed on 25 November 2018).
46. Anguelov, D.; Dulong, C.; Filip, D.; Frueh, C.; Lafon, S.; Lyon, R.; Ogale, A.; Vincent, L.; Weaver, J. Google Street View: Capturing the World at Street Level. *Computer* **2010**, *43*, 32–38. [CrossRef]

 *land*

Article

Assessing Stormwater Nutrient and Heavy Metal Plant Uptake in an Experimental Bioretention Pond

Giampaolo Zanin [1], Lucia Bortolini [2],* and Maurizio Borin [1]

[1] Department of Agronomy, Food, Natural Resources, Animals and Environment (DAFNAE), University of Padova, Viale dell'Università, 16, 35020 Legnaro, Italy; paolo.zanin@unipd.it (G.Z.); maurizio.borin@unipd.it (M.B.)

[2] Department of Land, Environment, Agriculture and Forestry (TESAF), University of Padova, Viale dell'Università 16, 35020 Legnaro, Italy

* Correspondence: lucia.bortolini@unipd.it; Tel +39-049-827-2735

Received: 29 October 2018; Accepted: 29 November 2018; Published: 1 December 2018

Abstract: With the purpose to study a solution based on Sustainable Urban Drainage Systems (SUDS) to reduce and treat stormwater runoff in urban areas, a bioretention pond (BP) was realized in the Agripolis campus of the University of Padova, Italy. The BP collected overflow water volumes of the rainwater drainage system of a 2270 m^2 drainage area consisting almost entirely of impervious surfaces. Sixty-six Tech-IA® floating elements, supporting four plants each, were laid on the water surface. Eleven species of herbaceous perennial helophyte plants, with ornamental features, were used and tested. The early growth results of the BP functioning showed that nearly 50% of the total inflow water volume was stored or evapotranspirated, reducing the peak discharge on the urban drainage system. Among plants, *Alisma parviflora*, *Caltha palustris*, *Iris* 'Black Gamecock', *Lysimachia punctata* 'Alexander', *Oenanthe javanica* 'Flamingo', *Mentha aquatica*, *Phalaris arundinacea* 'Picta', and *Typha laxmannii* had the best survival and growth performances. *A. parviflora* and *M. aquatica* appeared interesting also for pollutant reduction in runoff water.

Keywords: nature-based solution; floating treatment wetland; pollutant removal; runoff

---

1. Introduction

The high rate of urbanization has resulted in a large increase of impervious coverage in the landscape which can reach a very high percentage of the urban surface. Impervious surfaces decrease rainfall infiltration into the soil increasing runoff in terms of both peak flow and volume [1,2]. Rainwater in the urban landscape is therefore mainly directed into the municipal drainage system, creating serious problems in case of heavy rains, such as local floods, river inundations, etc., and reducing water availability and quality [3]. Urban runoff can be and often is a significant source of water pollution, causing a decline of fisheries, swimming areas, and other beneficial attributes of water resources [4]. At the same time, climate changes are causing the intensification and concentration of rainfall events, exacerbating the problem [5,6].

To reduce the problem, some environmentally sustainable approaches to urban development have been proposed as an alternative to the traditional ones to better manage the runoff in urban areas [7–10]. A stormwater best management practice (BMP) is a technique, measure, or structural control that is used to manage the quantity and improve the quality of stormwater runoff in the most cost-effective manner [11,12]. BMPs include stormwater planting, open channels, porous pavements, etc., in addition to a set of overall site design strategies and highly localized, small-scale, decentralized source control techniques, also known as Low-Impact Development (LID) systems in the USA [13,14] or Water-Sensitive Urban Design (WSUD) in Australia [15,16]. To describe stormwater technologies, such as bioretentions (including rain gardens), tree box filters, and green roofs, the term Sustainable

Urban Drainage Systems (SUDS) was also coined [12,17]. SUDS may be easily integrated into buildings, infrastructure, or landscape design, taking a decentralized approach to disperse flows and manage runoff closer to where it originates, rather than controlling it downstream in a large stormwater management facility [18–20]. Landscape designers have the opportunity to contribute to the mitigation of the stormwater management problem, by incorporating these solutions in the design of residential gardens, corporate and institutional landscapes, and public green spaces, in order to combine aesthetic quality objectives with functional gains for the development of a more sustainable landscape [21].

More recently [22], the term Blue-Green Infrastructure (BGI) has been used to define a planned network of natural and semi-natural areas that utilize natural processes to improve water quality and manage water quantity by restoring the hydrological function of the urban landscape and managing stormwater. In particular, bioretention structures are BGIs that mimic the hydrologic function of a natural landscape providing both flood control and water quality benefits [23].

An experimental project was conducted in the Agripolis Campus of the University of Padova (Italy) in order to evaluate the efficiency in runoff reduction and water quality improvement of two bioretention solutions characterized by different scale and slightly different functions.

One solution is a rain garden system, already investigated in other environmental conditions (e.g., [24–36]) but not in Italy, whose research results were recently published [37,38].

The other solution is a new proposal, i.e., a bioretention pond (BP) with impervious walls to store and treat stormwater runoff as in floating treatment wetland (FTW) systems [39,40] with living ornamental plants. The BP is intended for green areas within blocks, mall centers, etc., to create a setting with aesthetic features and also able to intercept and retain stormwater runoff, reducing the peak discharge into the drainage system or main stream network, decreasing pollutants in the overflow water, and eventually working as a water reservoir for sustainable supplemental irrigation of beddings or other plant settings during drought periods. Specifically, the objective of this paper was the evaluation of the capacity of the BP to manage stormwater runoff and of the plants response in terms of growth, aesthetic quality, and potential phytoremediation. The results related to the early growth period (first two vegetative seasons) are reported.

2. Materials and Methods

The bioretention pond was set in the Agripolis Campus of the University of Padova (Figure 1), in Legnaro (45°35′ N; 11°96′ E). The area has an annual average temperature of 12.3 °C and an average minimum and maximum temperatures of −5.5 and 32.8 °C. The average annual rainfall is 811 mm, mostly distributed during the growing season, from April to November.

During Spring 2011, in proximity of a building and a parking lot of the campus, a soil area of about 70 m^2 was dug up to a depth of about 165 cm. With the excavation, a storage basin of about 44.5 m^3 (8.10 × 5.23 m wide and 1.05 m deep) was obtained, and the entire basin was lined with a 1.5 mm thick polyolefin film.

The BP was designed to collect the overflow volumes of the existing rainwater drainage system of an area of 2270 m^2 consisting almost entirely of impervious surfaces (an asphalted road, sidewalks, and a building roof) (Figure 2). In fact, this drainage system has a good capacity to store and slowly let infiltrate into the soil all the runoff volumes from this area except during heavy or frequent rainfall events. The overflow volumes, which were previously discharged in a channel of the urban rainwater drainage system, were collected in a concrete sump and pumped into the BP. Water exceeding the storage capacity of the pond flowed into another sump from where it was pumped out in the channel of the local urban surface drainage system.

On July 2011, when the pond was almost full of water (102.5 mm in depth), 66 self-floating elements (Tech-IA®, Padova, Italy), supporting four plants each, were laid down on the water surface. Tech-IA® is a rectangular panel (0.50 × 0.90 m) produced in ethylene-vinil acetate (EVA), with eight gridded windows in which plants can be anchored. Its mass is 1732 g, and it may support a load capacity up to 20 kg [41,42]. The single elements were linked to each other, covering more than 70% of the storage basin surface (Figure 3).

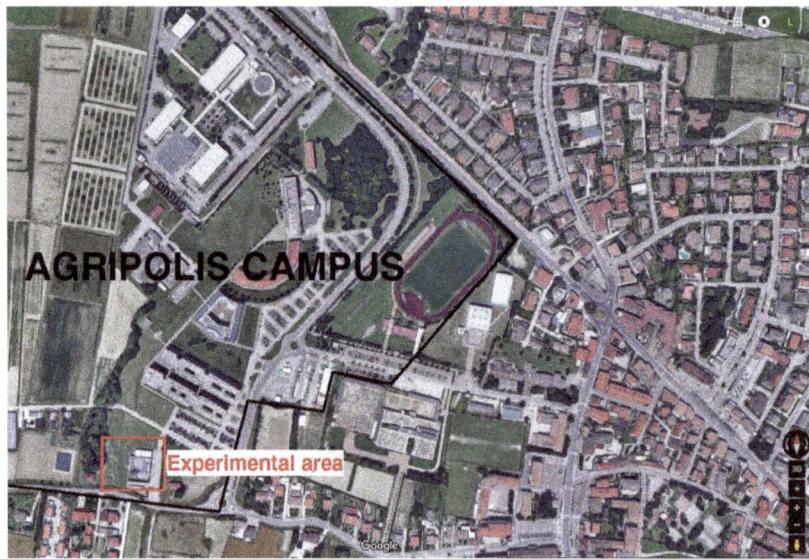

Figure 1. A Google Maps view of the Agripolis Campus of the University of Padova and the experimental area in which the two bioretention solutions (bioretention pond and rain garden system) are located.

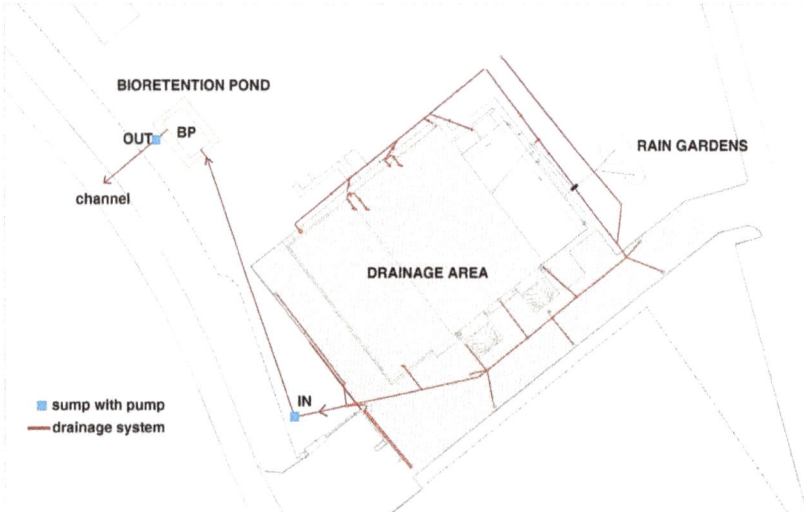

Figure 2. Plan of the experimental area with the position of the bioretention pond (BP) and the sumps where the water samples were collected (BP, IN, OUT). The rainwater drainage system with the path of water from storm drains is also reported. The overflow water was collected in the inflow sump (IN) and pumped in the BP; from the outflow sump (OUT), the water was pumped in a channel of the urban surface drainage system.

Figure 3. The bioretention pond at the end of the first growing season of the experiment.

The following 11 herbaceous perennial helophyte plants, with ornamental features, were used: *Alisma parviflora* Pursh., *Bacopa caroliniana* (Walt.) B.L. Robins, *Caltha palustris* L., *Canna indica* L., *Iris* 'Black Gamecock' (Louisiana Iris group), *Lysimachia punctata* L. 'Alexander', *Mentha aquatica* L., *Oenanthe javanica* Blume (DC.) 'Flamingo', *Phalaris arundinacea* L. 'Picta', *Sagittaria sagittifolia* L., and *Typha laxmannii* Lepech. (hereafter also called ALSSU, BAOCA, CTAPA, CNNIN, IRISS, LYSPU, MENAQ, OENJA, PHAAP, SAGSA and TYHLX, respectively according to their EPPO codes (see http://eppt.eppo.org/). For each species, six floating elements were adopted; the plants were set with the root system free in the water in three of these elements, while, in the others, the plants were set with roots confined in about 0.4 L of expanded clay, contained in plastic nets settled in place of the grids (Figure 4), through which roots could grow and reach the water.

Figure 4. Particular of two TECH-IA® elements in which the plants were set with (**above**) or without (**below**) the substrate of expanded clay.

Data collection considered the capacity of the BP in managing the overflow volumes from the existing rainwater drainage system subtracted to the canal of the urban drainage system. The inflow and outflow volumes were calculated on the basis of the operating time of the two drainage pumps (inflow pump and outflow pump) (Submersible pump MC/50-70, Pedrollo S.p.A., San Bonifacio, Italy), knowing their flow rate (1600 L min^{-1}).

The average daily actual evapotranspiration of the system was estimated in order to evaluate the capacity of the pond to ensure an adequate reservoir of water especially during dry periods. The values were calculated by measuring with a water level sensor (Levelogger Edge, Solinist Ltd., Georgetown, ON, Canada) the lowering of the water level in the pond during dry periods in different seasons. The values were compared with the average daily reference evapotranspiration ET_0 values calculated by the Penman–Monteith formula [43] using the data of the local weather station.

Samples of water were collected in the inflow and in the outflow sumps whenever the corresponding hour-counter revealed that pumps had operated. In these occasions, also three samples of water were collected from the pond at 20 cm of depth. Furthermore, samples of the BP water were collected every three weeks with no rainfall event. Water samples were analyzed for the concentration of nutrients and other ions (i.e., Cl^-, NO_3^-, PO_4^{3-}, SO_4^{2-}, Na^+, NH_4^+, K^+, Ca^+, and Mg^+), salinity, and dissolved heavy metals (i.e., Cu, Cb, Zn, and Pb). Nutrients were evaluated by means of ionic chromatography (ICS-900, Dionex, Sunnyvale, CA, USA); salinity, pH and heavy metal were determined with ICP SPECTRO CirOS Vision EOP (SPECTRO Analytical Instruments GmbH & Co., Kleve, Germany).

The plant characteristics at planting (i.e., height, leaf number, root length, dry weight) were determined in a sample of four plants per species. On November 2011, plant growth was evaluated by recording only in vivo parameters such as stem and leaf number, height, plant survival. The root growth was evaluated by means of its length and a visual rating (root visual rating RVR; 1–9 scale) based on root number and overall root growth. Furthermore, a visual rating was also adopted to evaluate the aesthetic values of the plants (aesthetic visual rating AVR; 1–9 scale) based on their potential growth in conventional condition. At the end of the second vegetative season (November 2012), the plants were evaluated as previously and, in addition, on a half of the plants, the dry weight of above-ground plant organs (AGPO) and below-ground plant organs (BGPO), comprehensive of rhizomes and stolons when present, was determined. For plants grown on substrate, the concentrations of Cd, Cu, Pb, and Zn in the dry matter were also determined, adopting [44] procedure for mineralization and ICP procedure for reading. The data were also used to calculate heavy metal content in both AGPO and BGPO, multiplying the concentrations by their respective dry weights.

The data were analyzed by mean of the analysis of variance. Statgraphics Centurion XVI software program (Statpoint Technologies, Inc., Warrenton, VA, USA) was used for data analysis. The data from the analysis of water collected in the BP were averaged before statistical analysis. The data on plant survival were analyzed by mean of the chi-square test. Non-linear regression (SigmaPlot for Windows 11.0; Systat Software, Inc., Chicago, IL, USA) was used to describe changes in nutrient concentration over time.

3. Results and Discussion

3.1. Hydrological Behaviour of the BP

During the period April 2011–November 2012, 121 rainfall events were recorded (a total of 944 mm of rain), but only 14 events generated overflow volumes from the rainwater drainage system. The total inflow volume in the BP was 245 m^3, and the water volume leaving the pond as outflow was 126 m^3, corresponding to 119 m^3 (nearly 50% of the total inflow volume) collected or evapotranspired by the BP system. This volume was therefore subtracted from the urban stormwater drainage system, reducing the peak flow rates in the canal during rainy periods. However, it is interesting to note that only 10% of the total potential runoff volume (about 2140 m^3 calculated by multiplying the drainage area for the rainfall) gave rise to overflow volume, because during the examined period the events were mostly of medium-low amount, perfectly managed by the existing rainwater drainage system.

In these occasions, the average daily evapotranspiration, calculated in no rainfall periods, was of 1.01 mm d^{-1} during wintertime (1 December 2011–28 February 2012), 3.03 mm d^{-1} in springtime

(1 April 2012–20 May 2012), and 3.32 mm d^{-1} in summertime (15 June 2012–25 August 2012). In the same periods, the average daily reference evapotranspiration ET_0 values calculated by the Penman–Monteith formula [43] were 0.95, 3.31, and 5.21 mm d^{-1}, respectively. The values of the actual evapotranspiration were relatively high if we consider the low transpiration of newly established plants: this was probably offset by the evaporation from the water surface left free by the floating elements (about 30% of the pond surface) and from the gridded windows without plants.

The BP was meant to guarantee a sufficient water depth for plant growth also in high-drought condition. However, even if rainfall events during the analyzed period were not frequent, the water depth in BP was high (over 100 cm deep) for most part of the experimental time. Only during summer 2012, the water level reached the lowest level (0.83 m) on 31 August. Nevertheless, as the average daily reference evapotranspiration ET_0 value during that summer was equal to 5.21 mm d^{-1}, considering a crop coefficient Kc equal to 1.2, as set for reed swamp [43] with a good plant growth, the actual ET of the system during the driest summer period could be up to 450 mm compared to the actual 285 mm that we observed. The hypothetical higher evapotranspiration would have resulted in lower depth of the storage water (about 0.60 m), which would have allowed the survival of the plants, demonstrating an adequate sizing of the BP.

3.2. Nutrient and Heavy Metal Concentration in Stormwater

The concentrations of nutrients and other ions in the BP inflow and outflow water were in general very low. Cl$^-$, Na$^+$, Mg$^+$, and Ca$^+$ concentrations in the BP water did not change over time (on average, 0.59, 12.4, 1.30, and 11.2 mg L^{-1}, respectively), and no difference between inflow and outflow water was found.

NO$_3^-$ concentration in the BP water was relatively high before plant establishment (1.53 mg L^{-1} on average until July 2011) and, after plant establishment, it was significantly lower (on average 0.355 mg L^{-1}). From April 2012, the values were sometimes lower than the limit of instrument detection (<0.02 mg L^{-1}). Figure 5, reporting the box and whisker diagrams of nitrate concentration in the BP inflow and outflow water throughout the experimental period, highlights that the outflow water had, as expected, the same concentration as in the BP water but much lower than in the inflow water.

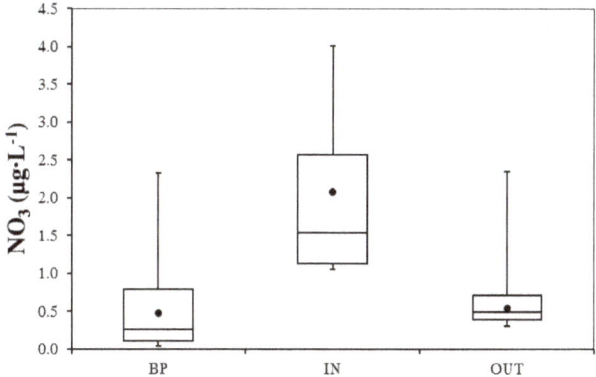

Figure 5. Concentration of nitrate (NO$_3$) in the BP (n = 18 samples), inflow water (IN) (n = 14 samples) and outflow water (OUT) (n = 6 samples). Each box shows the median and range between first and third quartile of all samples, while the whiskers show the minimum and maximum values. The mean (●) is also shown.

PO$_4^{3-}$ concentration in water samples collected during the first growing season was low, on average 1.22 mg L^{-1}, and not always detected. In the second growing season, all samples had lower concentration values than detectable (actual sensibility of the instrument >1.0 mg L^{-1}). In contrast to what reported for NO_3^-, no differences were noted among inflow, outflow, and BP water samples. SO_4^{2-}, NH_4^+, and K^+ concentrations had the same pattern described for NO_3^-, but differences resulted significant only for K^+. As reported in Figure 6, K^+ concentrations in inflow water were higher than those in the BP, while outflow water had intermediate values. These low values of nutrients and other ions are more comparable with those of rainwater and runoff from roofs than with those of the runoff from trafficked areas [45].

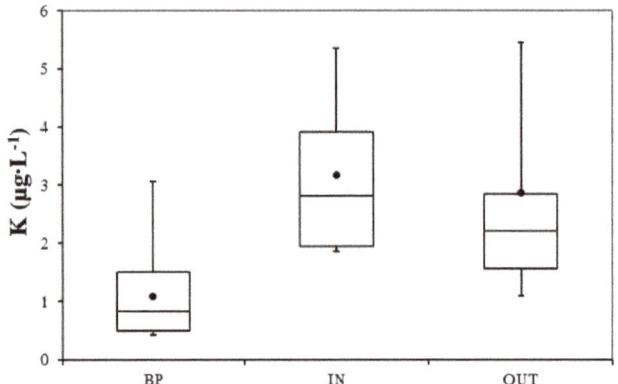

Figure 6. Concentration of potassium (K) in the BP (n = 18 samples), IN (n = 14 samples) and OUT (n = 6 samples) waters. Each box shows the median and range between first and third quartile of all samples, while the whiskers show the minimum and maximum values. The mean (•) is also shown.

Salinity takes into account the presence of all ions in water. Salinity values did not change over time and not even differed in inflow and outflow waters (ranging from 44.7 to 89.0 μS cm^{-1} in BP water, from 44.0 to 115 μS cm^{-1} in inflow water, and from 54.0 to 89.0 μS cm^{-1} in outflow water).

The water pH also did not change over time, and the values (ranging from 7.02 to 7.62) were similar to those of rainwater [45,46].

The concentrations of dissolved heavy metals in the water were in general very low and, in the case of Cd and Pb, the values were below the sensibility of the instrument (<0.001 and <0.005 mg L^{-1}, respectively). Concentrations of Cu and Zn did not apparently change over time or in inflow or outflow waters and averaged 0.007 mg L^{-1} (range 0.002–0.012 mg L^{-1}) and 0.004 mg L^{-1} (range 0.001–0.028 mg L^{-1}), respectively. As seen for ions, the concentrations of heavy metals were more comparable with those of rainwater and runoff from roofs than with those of the runoff from trafficked areas [45,47] or at the outlet of storm sewers [46].

The FTWs have proven to be efficient in ameliorating stormwater quality both at mesocorm experiment level [48] and at field level [41,42,49,50], leading to improvements as high as 14% for NO_3–N, 65–75% for Cu, and 40% for Zn after seven days of treatment. In the present study, the apparently low or no effect was probably due to the low concentrations of both nutrients and heavy metals in the inlet water, little surface coverage, and growth stage of the plants.

3.3. Plant Growth and Heavy Metal Accumulation

The characteristics of the plants at planting and their performance at the end of the first and second year are reported in Tables 1–4. *A. parviflora* (ALSSU) plant material used for transplant arrived from a traditional cultivation in soil. Nevertheless, all plants were alive at the end of the first year, and a small growth occurred, without any difference among treatments. At the end of the experiment, differences were found only for the biomass of AGPO and BGPO and, of course, of the whole plant (WP) (Table 1). The presence of the substrate improved ALSSU growth and, in fact, the related values were almost doubled (Table 1).

The plant material utilized for the transplant of *B. caroliniana* (BAOCA) was merely stem cutting with preformed roots. The presence of a substrate facilitated plant establishment. Furthermore, the plants had 90% more developed buds and a better overall appearance (Table 2). The plants of BAOCA grown without substrate died during winter, while two-thirds of plants cultivated in the substrate remained alive. However, the biomass data collected at the end of the second year indicate that growth was very poor (Table 1).

Table 1. Dry weight of above-ground plant organs (AGPO), below-ground plant organs (BGPO), and whole plant (WP) at the beginning and at the end of the experiment (g plant^{-1} ± sd).

	Beginning of the Experiment	End of the Experiment			Beginning of the Experiment	End of the Experiment		
		Substrate				Substrate		
		No	Yes	Sig ^		No	Yes	Sig ^
	Alisma parviflora (ALSSU)				*Mentha aquatic* (MENAQ)			
AGPO	5.37 ± 1.81	2.96	6.78	*	4.66 ± 2.12	12.8	38.6	*
BGPO	0.801 ± 0.22	2.50	4.52	*	0.633 ± 0.22	23	47.1	*
WP	6.17 ± 1.51	5.46	11.3	*	5.29 ± 2.21	35.8	85.7	*
	Bacopa caroliniana (BAOCA)				*Oenanthe javanica* 'Flamingo' (OENJA)			
AGPO	0.257 ± 0.22	-	0.288		3.13 ± 0.37	-	6.32	
BGPO	1.36 ± 0.32	-	0.260		0.290 ± 0.35	-	11.4	
WP	1.62 ± 0.53	-	0.548		3.42 ± 0.36	-	17.7	
	Canna indica (CNNIN)				*Phalaris arundinacea* 'Picta' (PHAAP)			
AGPO	1.49 ± 0.57				3.05 ± 0.12	0.80	3.68	*
BGPO	1.11 ± 0.21				1.75 ± 0.14	1.32	8.77	***
WP	2.60 ± 0.66				4.80 ± 0.23	2.12	12.4	**
	Caltha palustris (CTAPA)				*Sagittaria sagittifolia* (SAGSA)			
AGPO	4.79 ± 1.00	5.77	11.41	ns	0.501 ± 0.21	-	0.297	
BGPO	9.67 ± 3.51	13.4	31.4	*	0.440 ± 0.15	-	0.548	
WP	14.5 ± 3.83	19.1	42.8	*	0.941 ± 0.43	-	0.844	
	Iris 'Black Gamecock' (IRISS)				*Typha laxmannii* (TYHLX)			
AGPO	0.91 ± 0.19	2.34	3.65	ns	1.26 ± 0.35	4.13	11.8	*
BGPO	2.91 ± 0.96	9.38	14.8	ns	1.93 ± 1.20	2.07	11.7	*
WP	3.82 ± 1.14	11.7	18.4	ns	3.19 ± 1.55	6.2	23.5	**
	Lysimachia punctata 'Alexander' (LYSPU)							
AGPO	1.94 ± 1.57	8.24	11.3	ns				
BGPO	1.05 ± 0.94	7.18	10.9	ns				
WP	2.99 ± 0.63	15.4	22.2	ns				

^ ***, ** and *: significant at $p \leq 0.001$, 0.01 and 0.05, respectively. ns = non-significant.

All plants of *C. palustris* (CTAPA) survived and, at the end of the first year, the plants grown with substrate showed better parameters, with the only exception of root length (Table 2). At the end of the experiment, the improvement shown by CTAPA grown with substrate was less evident (Tables 1 and 2).

Rhizome cuttings with shoot and poor root system were used for *C. indica* (CNNIN). At the end of the first year, the plants grown with substrate differed from those without it only in the root system features: the former had shorter but more numerous roots (Table 2). Despite the American Horticultural Society considers CNNIN quite hardy for our winter temperature (www.ahs.org), the plants did not survive through the winter (minimum temperature registered −7.4 °C).

Table 2. Plants characteristics at the beginning of the experiment, at the end of the first year, and at the end of the second year (RVR = root visual score; AVR = aesthetic visual score).

	Beginning of the Experiment	End of the First Year			End of the Second Year		
		Substrate			Substrate		
		No	Yes	Sig ^	No	Yes	Sig ^
A. parviflora (ALSSU)							
Stem number	1.00 ± 0.0	1.13	1.13	ns	2.27	1.75	ns
Plant height (cm)	22.4 ± 4.87	24	20.2	ns	11.6	13.4	ns
Leaf number	4.75 ± 1.86	6.1	7.2	ns	20.3	17.7	ns
Root length (cm)	3.92 ± 1.16	41.3	46.4	ns	70.4	63.2	ns
RVR (1–9 scale)		1.5	1.67	ns	4.36	2.75	ns
AVR (1–9 scale)		2.08	1.75	ns	3	2.42	ns
Mortality (%)					8.33	0.0	ns
B. caroliniana (BAOCA)							
Stem number	5.96 ± 1.46	3.75	3.92	ns	-	5.25	
Plant height (cm)	15.7 ± 4.6	8.08	8.92	ns	-	6.63	
Root length (cm)	6.50 ± 1.93	25.0	15.3	**	-	16.1	
Bud number		7.00	13.3	***			
RVR (1–9 scale)		2.08	2.58	ns	-	1.38	
AVR (1–9 scale)		2.58	5.10	**	-	1.38	
Survival (%)		100	100	ns	0.0	66.7	***
C. indica (CNNIN)							
Stem number	1.00 ± 0.0	1.17	1.67	ns			
Plant height (cm)	27.9 ± 6.96	21.2	24.3	ns			
Leaf number	4.17 ± 1.19	6.67	8.83	ns			
Root length (cm)	6.33 ± 1.03	67.5	48.8	*			
RVR (1–9 scale)		2.00	3.67	*			
AVR (1–9 scale)		2.50	3.50	ns			
Survival (%)		100	100	ns	0.0	0.0	ns
C. palustris (CTAPA)							
Stem number	3.10 ± 1.41	2.33	3.58	**	4.50	7.75	ns
Plant height (cm)	14.2 ± 1.56	4.58	11.0	***	30.5	65.5	*
Leaf number	4.2 ± 1.92	2.50	6.58	***	10.2	16.9	ns
Root length (cm)	11.7 ± 1.10	28.7	20.7	***	50.9	68.6	*
RVR (1–9 scale)		3.50	4.50	*	4.40	7.75	ns
AVR (1–9 scale)		1.92	4.08	**	3.20	5.25	*
Survival (%)		100	100	ns	83.3	100	*

^ ***, ** and *: significant at $p \leq 0.001, 0.01$ and 0.05, respectively. ns = non-significant.

The plant material used for *Iris* (IRISS) and *L. punctata* (LYSPU) transplant was not adapted to the floating system, but no death was observed for both species (Table 3). Some parameters were significantly higher in plants grown with substrate (e.g., plant height of IRISS and stem number of LYSPU), but no differences were observed in biomass accumulation (Table 1).

The plants of *M. aquatica* (MENAQ) and *P. arundinacea* (PHAAP) showed a better growth with substrate already at the end of the first year, with significant higher values for all the observed parameters (see Tables 1, 3 and 4). These species responded similarly also in plant survival, with 75 and a 50% of death for the two species, occurred only in absence of substrate.

Plants of *O. javanica* (OENJA) had also better performance in the presence of substrate (Table 3), and, at the end of the experiment, only the plants grown with the substrate were still alive.

The same behavior was observed for *S. sagittifolia* (SAGSA) and *T. laxmannii* (THYLX) plants. During the first year, a 16.7% of SAGSA plants growth without substrate died, and the remaining did not survive winter. However, a very poor growth was observed also for plants grown in the substrate (Tables 1 and 4). Regarding TYHLX, 91.7% of plants grown with the substrate survived, while only 58.3% of those grown without the substrate died (Table 4). The growth of the remaining plants was greatly improved by the substrate and, with the only exception of root length, all parameters showed values increased by over 100% (Tables 1 and 4).

Table 3. Plants characteristics at the beginning of the experiment, at the end of the first year, and at the end of the second year.

	Beginning of the Experiment	End of the First Year			End of the Second Year		
		Substrate			Substrate		
		No	Yes	Sig ^	No	Yes	Sig ^
Iris 'Black Gamecock' (IRISS)							
Stem number	1.08 ± 0.28	1.92	1.67	ns	2.30	2.42	***
Plant height (cm)	29.1 ± 4.02	13.0	18.3	*	32.6	36.3	**
Leaf number	4.25 ± 1.54	11.8	11.1	ns	15.3	16.7	ns
Root length (cm)	5.25 ± 0.89	23.1	29.3	**	33.7	38.0	ns
RVR (1–9 scale)		2.33	5.25	***	2.30	2.75	*
AVR (1–9 scale)		1.92	2.17	ns	2.83	2.58	**
Survival (%)		100	100	ns	100	100	ns
L. punctata 'Alexander' (LYSPU)							
Stem number	1.79 ± 0.72	2.17	3.75	*	4.00	7.83	*
Plant height (cm)	33.7 ± 6.05	7.33	8.17	ns	23.3	26.3	ns
Root length (cm)	4.67 ± 1.23	24.8	21.5	ns	48.8	52.5	ns
New shoot number		0.83	1.83	ns			
RVR (1–9 scale)		2.58	3.08	ns	3.50	4.58	ns
AVR (1–9 scale)		2.42	3.33	ns	3.00	4.75	*
Survival (%)		100	100	ns	100	100	ns
Mentha aquatica (MENAQ)							
Stem number	6.54 ± 2.08	8.42	9.00	ns	5.67	23.8	*
Plant height (cm)	24.7 ± 3.62	36.0	39.5	ns	38.0	64.3	***
Root length (cm)	6.31 ± 4.25	30.4	37.7	*	79.4	100.8	**
New shoot number		0.83	3.50	**			
RVR (1–9 scale)		7.75	8.17	ns	5.78	8.33	*
AVR (1–9 scale)		3.25	5.08	**	4.11	7.67	*
Survival (%)		100	100	ns	75	100	**
Oenanthe javanica 'Flamingo' (OENJA)							
Stem number	5.88 ± 1.08	7.5	16.1	***	-	24.8	
Plant height (cm)	29.4 ± 4.05	8.58	10.2	*	-	25.6	
Root length (cm)	6.17 ± 2.04	20.8	35.8	***	-	57.9	
New shoot number		1.92	9.92	***			
RVR (1–9 scale)		2.33	7.58	***	-	6.67	
AVR (1–9 scale)		2.42	4.92	**	-	3.92	
Survival (%)		100	100	ns	0.0	100	***

^ ***, ** and *: significant at $p \leq 0.001$, 0.01 and 0.05, respectively, ns = non-significant.

Table 4. Plants characteristics at the beginning of the experiment, at the end of the first year, and at the end of the second year.

	Beginning of the Experiment	End of the First Year			End of the Second Year		
		Substrate			Substrate		
		No	Yes	Sig ˆ	No	Yes	Sig ˆ
Phalaris arundinacea 'Picta' (PHAAP)							
Stem number	1.54 ± 0.59	5.00	7.08	*	2.50	9.92	*
Plant height (cm)	45.8 ± 13.9	21.7	21.1	ns	14.8	31.4	*
Root length (cm)	6.00 ± 2.17	37.5	35.8	ns	20.2	38.8	*
RVR (1–9 scale)		4.83	5.92	*	2.00	3.75	*
AVR (1–9 scale)		2.75	4.67	**	1.33	3.92	*
Survival (%)		100	100	ns	50	100	***
Sagittaria sagittifolia (SAGSA)							
Stem number	1.00 ± 0.0	1.40	1.00	ns	-	1.09	
Plant heigh (cm)	12.3 ± 2.07	13.2	15.6	*	-	14.1	
Leaf number	3.91 ± 2.66	5.13	7.14	*	-	16.6	
Root length (cm)	7.00 ± 3.30	32.0	33.1	ns	-	42.6	
RVR (1–9 scale)		2.70	2.58	ns	-	1.64	
AVR (1–9 scale)		2.42	3.33	*	-	1.55	
Survival (%)		83.3	100	*	0.0	91.7	***
Typha laxmannii (TYHLX)							
Stem number	1.13 ± 0.34	1.75	2.92	*	2.14	4.82	*
Plant height (cm)	44.3 ± 3.68	14.8	35.2	ns	46.3	62.6	*
Leaf number	5.83 ± 1.63	7.17	13.58	***	16.1	37.2	*
Root length (cm)	3.44 ± 2.22	12.3	14.9	ns	33.6	45.5	ns
RVR (1–9 scale)		1.25	2.50	**	2.14	4.36	*
AVR (1–9 scale)		1.92	3.17	**	1.57	4.00	**
Survival (%)		100	100	ns	58.3	91.7	**

ˆ ***, ** and *: significant at $p \leq 0.001, 0.01$ and 0.05, respectively, ns = non-significant.

From these results, it is clear that both survival and growth of plants in the Tech_IA® elements were promoted by the presence of a substrate (Figure 7). This was probably due to a better root environment (e.g., humidity around the collar point) which initially favoured rooting and promoted plant establishment. Apart from that, plants like SAGSA, with clumping habit and weak structures, need a suitable anchorage for their growth. On the contrary, the attitude to produce stolons or rhizomes (e.g., *Mentha* and *Iris*) favours anchorage to Tech-IA® elements.

(a)　　　　　　　　　　　(b)

Figure 7. Different plant growth of *M. aquatica* with (**a**) or without (**b**) the substrate.

When evaluating species survival to winter, it appeared that *C. indica* is not useful in our environment, as well as *Bacopa* and *Sagittaria*, whose poor growth indicates their poor adaptability to the employed BP system.

In order to evaluate the ability of the selected plants to improve water quality, the accumulation of heavy metal was also measured. The evaluation considered only the species with good growth results and high plant survival, cultivated with the substrate. As reported in Table 3, the heavy metal with the highest concentration was Zn followed by Cu, Pb, and Cd. Furthermore, in general, higher concentrations were found in the AGPO than in the BGPO.

Among plants species, *A. parviflora* had the highest values of all heavy metals. On the contrary, *Iris* had the lowest values of Cu, Pb, and Zn. *Mentha*, *Phalaris*, and *Typha* had the overall lowest values of Cd. The concentration of Zn found in this research is higher than those found by another study [51] in *P. arundinacea* and other species grown in normal condition, or comparable with those found in plants grown with nutrient solutions. Furthermore, according to reference [52], the concentrations of heavy metals are within normal levels even if the values found in water are relatively low.

If we consider the heavy metals accumulated on a mass basis, the highest values were found in *C. palustris* and *M. aquatica*. It is worth noting that ALSSU, as well as MENAQ, CTAPA, LYSPU, and OENJA had a good accumulation of these elements in the AGPO (Table 5).

As the management of the BP includes an annual cleaning of Tech-IA elements with the removal of the aerial part of plants, the content of heavy metals in AGPO can be the most important factor to consider if the plant selection is done on the basis of its phytoremediation ability. In fact, the heavy metals are removed with the removal of the aerial part of plants, while they remain in BGPO and, because of the decay of the root system, they can be released in the water of the pond.

A last consideration has to be made. In a sustainable approach to manage storm water in an urban context, no nutritional elements were provided to the plants. As the water arriving at the pond was very poor in nutrients, plant growth of all species was poorer than in normal nutritional conditions, as evidenced by the nutritional deficiency symptoms that were observed in all species (Figure 8). It is possible that, if a controlled-release fertilizer was applied on the substrate during active plant growth (i.e., in springtime), the surface growth and aesthetic appearance of the plants could be improved.

(a)   (b)

Figure 8. Nutritional deficiency symptoms were observed in all species: the example of *Iris* cultivated with (**a**) and without (**b**) substrate is reported.

Table 5. Heavy metal concentration and content (mean ± standard deviation) in the dry matter of AGPO, BGPO, and WP of the species cultivated with the substrate.

	Heavy Metal Concentration (μg g^{-1} Dry Matter)				Heavy Metal Content (μg Plant^{-1})			
	Cd	Cu	Pb	Zn	Cd	Cu	Pb	Zn
Alisma parviflora (ALSSU)								
AGPO	0.454 ± 0.338	51.5 ± 19.9	1.55 ± 1.80	309 ± 108	2.09 ± 0.61	288 ± 94	6.2 ± 5.4	1739 ± 568
BGPO	0.490 ± 0.270	97.6 ± 38.6	9.97 ± 6.04	413 ±130	1.56 ± 0.34	359 ± 160	30.6 ± 5.9	1553 ± 745
WP					3.66 ± 0.28	647 ± 244	36.8 ± 2.8	3292 ± 1053
Caltha palustris (CTAPA)								
AGPO	0.076 ± 0.026	29.6 ± 3.1	0.60 ± 0.01	105 ± 28	0.83 ± 0.20	339 ± 103	6.8 ± 1.7	1163 ± 220
BGPO	0.330 ± 0.049	70.2 ± 4.5	2.54 ± 0.61	248 ± 24	10.32 ± 1.18	2199 ± 58	80.0 ± 20.5	7761 ± 373
WP					11.15 ± 1.00	2538 ± 54	86.9 ± 19.2	8924 ± 562
Iris 'Black Gamecock' (IRISS)								
AGPO	0.269 ± 0.119	14.0 ± 1.6	0.37 ± 0.14	102 ± 14	0.93 ± 0.21	51 ± 8	1.4 ± 0.6	372 ± 99
BGPO	0.120 ± 0.026	16.9 ± 2.5	0.63 ± 0.56	59 ± 10	1.74 ± 0.59	254 ± 100	10.3 ± 10.1	902 ± 395
WP					2.67 ± 0.40	304 ± 107	11.7 ± 10.7	1274 ± 480
Lysimachia punctata 'Alexander' (LYSPU)								
AGPO	0.168 ± 0.113	20.4 ± 6.1	1.27 ± 0.99	94 ± 16	1.01 ± 0.38	206 ± 86	10.1 ± 2.2	1060 ± 662
BGPO	0.229 ± 0.160	51.9 ± 3.1	2.87 ± 0.74	182 ± 38	2.48 ± 2.45	562 ± 556	30.8 ± 31.0	1959 ± 1964
WP					3.49 ± 2.13	768 ± 585	40.9 ± 33.1	3019 ± 2066
Mentha aquatica (MENAQ)								
AGPO	0.060 ± 0.026	12.6 ± 1.3	0.21 ± 0.07	44 ± 11	2.58 ± 2.22	495 ± 254	8.5 ± 6.0	1787 ± 1219
BGPO	0.154 ± 0.106	41.3 ± 7.6	2.15 ± 0.40	133 ± 18	7.03 ± 4.54	2023 ± 1042	104.1 ± 48.9	6193 ± 1898
WP					9.61 ± 4.89	2518 ± 1285	112.6 ± 53.8	7980 ± 2958
Oenanthe javanica 'Flamingo' (OENJA)								
AGPO	0.121 ± 0.027	31.0 ± 1.1	0.44 ± 0.18	193 ± 10	0.79 ± 0.40	197 ± 75	3.0 ± 2.3	1208 ± 386
BGPO	0.481 ± 0.133	57.4 ± 4.9	2.34 ± 0.75	339 ± 57	4.97 ± 1.86	637 ± 304	25.9 ± 17.0	3649 ± 1541
WP					5.76 ± 1.66	834 ± 240	28.9 ± 15.6	4857 ± 1196
Phalaris arundinacea 'Picta' (PHAAP)								
AGPO	0.150 ± 0.105	21.6 ± 2.4	0.51 ±0.15	220 ± 18	0.55 ± 0.38	79 ± 5	1.9 ± 0.5	807 ± 43
BGPO	0.077 ± 0.051	61.7 ± 7.1	1.66 ± 0.12	270 ± 44	0.67 ± 0.42	540 ± 51	14.6 ± 1.1	2357 ± 325
WP					1.22 ± 0.59	619 ± 56	16.5 ± 0.8	3164 ± 346
Typha laxmannii (TYHLX)								
AGPO	0.075 ± 0.052	13.9 ± 1.0	0.27 ± 0.12	62 ± 7	0.92 ± 0.72	164 ± 25	3.2 ± 1.5	741 ± 156
BGPO	0.146 ± 0.006	48.1 ± 8.9	3.27 ± 1.14	190 ± 35	1.71 ± 0.33	571 ±186	39.8 ± 20.2	2194 ± 343
WP					2.63 ± 0.75	735 ± 208	42.9 ± 21.6	2935 ± 496

4. Conclusions

The results of the early growth period demonstrate that the BP system can be an interesting approach, among the SUDS solutions, to increase sustainable stormwater management in urban areas, because of its capacity to storage runoff volumes (encouraging alternative uses such as irrigation of flower beds) and to subtract them to the urban drainage system, ths reducing the peak discharge during heavy rainfall periods.

Some of the evaluated species (i.e., *A. parviflora*, *C. palustris*, *Iris* 'Black Gamecock', *L. punctata* 'Alexander', *O. javanica* 'Flamingo', *M. aquatica*, *P.arundinacea* 'Picta', and *T. laxmannii*) seem to be adaptable to this particular growing system, especially if a substrate is adopted. In particular, the highest biomass production was obtained with *M. aquatica* and *C. palustris*, with 85.7 and 42.8 g plant^{-1} dry weight and 7.67 and 5.25 aesthetic visual score, respectively. *A. parviflora* appeared interesting for heavy metal concentration in plant tissue, but the higher biomass production makes *M. aquatica* and *C. palustris* interesting for pollutant reduction (e.g., 2.5 and about 8.0 mg plant^{-1} of Cu and Zn for both species) of stormwater as well.

Further research is needed to evaluate the opportunity to add slow-release nutrients to improve plant growth and appearance in order to obtain an aesthetically and hydrologically functional green infrastructure for urban landscapes, also reducing pollutant loads.

Author Contributions: Conceptualization, G.Z., L.B., and M.B.; Data curation, G.Z. and L.B.; Formal analysis, G.Z., L.B., and M.B.; Funding acquisition, G.Z.; Investigation, L.B. and G.Z.; Methodology, G.Z., L.B., and M.B.; Writing—original draft, G.Z., L.B., and M.B.; Writing—review & editing, L.B.

Funding: This research was funded by University of Padova—Progetto di Ateneo 2009 N. CPDA099823 entitled "Green structures for runoff control in urban environments".

Conflicts of Interest: The authors declare no conflict of interest.

References

1. Jennings, D.B.; Jarnagin, S.T. Changes in anthropogenic impervious surfaces, precipitation and daily streamflow discharge: A historical perspective in a mid-Atlantic subwatershed. *Landscape Ecol.* **2002**, *17*, 471–489. [CrossRef]
2. Dietz, M.E.; Clausen, J.C. Stormwater runoff and export changes with development in a traditional and low impact subdivision. *J. Environ. Manag.* **2008**, *87*, 560–566. [CrossRef] [PubMed]
3. Deletic, A.; Maksumovic, C.T. Evaluation of water quality factors in storm runoff from paved areas. *J. Environ. Eng.* **1998**, *124*, 869–879. [CrossRef]
4. US EPA. *Handbook—Urban Runoff Pollution Prevention and Control Planning*; EPA 625-R-93-004; U.S. Environmental Protection Agency: Washington, DC, USA, 1993.
5. Bronstert, A.; Niehoff, D.; Bürger, G. Effects of climate and land-use change on storm runoff generation: Present knowledge and modelling capabilities. *Hydrol. Process.* **2002**, *16*, 509–529. [CrossRef]
6. Sofia, G.; Roder, G.; Dalla Fontana, G.; Tarolli, P. Flood dynamics in urbanised landscapes: 100 years of climate and humans' interaction. *Sci. Rep.* **2017**, *7*, 40527. [CrossRef] [PubMed]
7. D'Arcy, B.; Frost, A. The role of best management practices in alleviating water quality problems associated with diffuse pollution. *Sci. Total Environ.* **2001**, *265*, 359–367. [CrossRef]
8. Lloyd, S.D. *Water Sensitive Urban Design in the Australian Context*; Technical Report No. 01/7; Cooperative Research Centre for Catchment Hydrology: Melbourne, Australia, 2001.
9. Prince George's County. *Bioretention Manual*; PGC: Landover, MD, USA, 2007.
10. Woods-Ballard, B.; Kellagher, R.; Martin, P.; Jefferies, C.; Bray, R.; Shaffer, P. *The SUDS Manual*; CIRIA C697: London, UK, 2007.
11. Field, R.; Tafuri, A.N.; Muthukrishnan, S.; Acquisto, B.A.; Selvakumar, A. (Eds.) *The Use of Best Management Practices (BMPs) in Urban Watersheds*; DEStech Publications Inc.: Lancaster, PA, USA, 2006.
12. Fletcher, T.D.; Shuster, W.; Hunt, W.F.; Ashley, R.; Butler, D.; Arthur, S.; Trowsdale, S.; Barraud, S.; Semadeni-Davies, A.; Bertrand-Krajewski, J.; et al. SUDS, LID, BMPs, WSUD and more–The evolution and application of terminology surrounding urban drainage. *Urban Water J.* **2015**, *12*, 525–542. [CrossRef]
13. Prince George's County. *Low-Impact Development Design Strategies: An Integrated Design Approach*; MD Department of Environmental Resources: Largo, MD, USA, 2000.
14. US EPA. *National Management Measures to Control Nonpoint Source Pollution from Urban Areas*; EPA 841-B-05-004; U.S. Environmental Protection Agency: Washington, DC, USA, 2005.
15. Coutts, A.M.; Tapper, N.J.; Beringer, J.; Loughnan, M.; Demuzere, M. Watering our cities: The capacity for Water Sensitive Urban Design to support urban cooling and improve human thermal comfort in the Australian context. *Prog. Phys. Geogr.* **2013**, *37*, 2–28. [CrossRef]
16. Melbourne Water. *WSUD Engineering Procedures: Stormwater*; CSIRO Publishing: Melbourne, Australia, 2005.
17. CIRIA. *Sustainable Urban Drainage Systems: Design Manual for Scotland and Northern Ireland*; CIRIA C521; Construction Industry Research and Information Association: London, UK, 2000.
18. Dietz, M.E. Low impact development practices: A review of current research and recommendations for future directions. *Water Air Soil Pollut.* **2007**, *186*, 351–363. [CrossRef]
19. Chang, N.B. Hydrological connections between low-impact development, watershed best management practices, and sustainable development. *J. Hydrol. Eng.* **2010**, *15*, 384–385. [CrossRef]
20. Roy-Poirier, A.; Champagne, P.; Filion, Y. Review of bioretention system research and design: Past, present, and future. *J. Environ. Eng.* **2010**, *136*, 878–889. [CrossRef]
21. Bortolini, L.; Semenzato, P. Low impact development techniques for urban sustainable design: A rain garden case study. *Acta Hortic.* **2010**, *881*, 327–330. [CrossRef]
22. Brears, R.C. Blue-Green Infrastructure in Managing Urban Water Resources. In *Blue and Green Cities*; Palgrave Macmillan Ed.: London, UK, 2018; pp. 43–61.
23. DeBusk, K.M.; Hunt, W.F.; Line, D.E. Bioretention outflow: Does it mimic nonurban watershed shallow interflow? *J. Hydrol. Eng.* **2010**, *16*, 274–279. [CrossRef]
24. Dietz, M.E.; Clausen, J.C. A field evaluation of rain garden flow and pollutant treatment. *Water Air Soil Pollut.* **2005**, *167*, 123–138. [CrossRef]

25. Hunt, W.F.; Jarrett, A.R.; Smith, J.T.; Sharkey, J.L. Evaluating bioretention hydrology and nutrient removal at three field sites in North Carolina. *J. Irrig. Drain. Eng.* **2006**, *132*, 600–608. [CrossRef]

26. Muthanna, T.M.; Viklander, M.; Blecken, G.T.; Thorolfsson, S.T. Snowmelt pollutant removal in bioretention areas. *Water Res.* **2007**, *41*, 4061–4072. [CrossRef] [PubMed]

27. Davis, A.P. Field performance of bioretention: Hydrology impacts. *J. Hydrol. Eng.* **2008**, *13*, 90–95. [CrossRef]

28. Stander, E.K.; Borst, M.; O'Connor, T.P.; Rowe, A.A. The effects of rain garden size on hydrological performance. In Proceedings of the IECA Northeast Chapter Conference and Trade Show, Hartford, CT, USA, 27–29 October 2009.

29. Guo, J.C.Y. Cap-orifice as a flow regulator for rain garden design. *J. Irrig. Drain. Eng.* **2012**, *138*, 198–202. [CrossRef]

30. Sidek, L.M.; Muha, N.E.; Noor, N.A.M.; Basri, H. Constructed rain garden systems for stormwater quality control under tropical climates. *IOP Conf. Ser. Earth Environ. Sci.* **2013**, *16*, 012020. [CrossRef]

31. Yergeau, S.E.; Obropta, C.C. Preliminary field evaluation of soil compaction in rain gardens. *J. Environ. Eng.* **2013**, *139*, 625–634. [CrossRef]

32. Turk, R.L.; Kraus, H.T.; Bilderback, T.E.; Hunt, W.F.; Fonteno, W.C. Rain garden filter bed substrates affect stormwater nutrient remediation. *HortScience* **2014**, *49*, 645–652.

33. Richards, P.J.; Farrell, C.; Tom, M.; Williams, N.S.; Fletcher, T.D. Vegetable raingardens can produce food and reduce stormwater runoff. *Urban For. Urban Green.* **2015**, *14*, 646–654. [CrossRef]

34. Mehring, A.S.; Hatt, B.E.; Kraikittikun, D.; Orelo, B.D.; Rippy, M.A.; Grant, S.B.; Gonzales, J.P.; Jiang, S.C.; Ambrose, R.F.; Levin, L.A. Soil invertebrates in Australian rain gardens and their potential roles in storage and processing of nitrogen. *Ecol. Eng.* **2016**, *97*, 138–143. [CrossRef]

35. Tang, S.; Luo, W.; Jia, Z.; Liu, W.; Li, S.; Wu, Y. Evaluating retention capacity of infiltration rain gardens and their potential effect on urban stormwater management in the Sub-humid Loess Region of China. *Water Res. Manag.* **2016**, *30*, 983–1000. [CrossRef]

36. Ishimatsu, K.; Ito, K.; Mitani, Y.; Tanaka, Y.; Sugahara, T.; Naka, Y. Use of rain gardens for stormwater management in urban design and planning. *Landsc. Ecol. Eng.* **2017**, *13*, 205–212. [CrossRef]

37. Bortolini, L.; Zanin, G. The experimental and educational rain gardens of the Agripolis Campus (north-east Italy): Preliminary results on hydrological and plant behavior. *Acta Hortic.* **2017**, *1189*, 531–536. [CrossRef]

38. Bortolini, L.; Zanin, G. Hydrological behaviour of rain gardens and plant suitability: A study in the Veneto plain (north-eastern Italy) conditions. *Urban For. Urban Green* **2018**, *34*, 121–133. [CrossRef]

39. Headley, T.R.; Tanner, C.C. *Applications of Floating Wetlands for Enhanced Stormwater Treatment: A Review*; Auckland Regional Council, Technical Publication: Auckland, New Zealand, 2006.

40. Ge, Z.; Feng, C.; Wang, X.; Zhang, J. Seasonal applicability of three vegetation constructed floating treatment wetlands for nutrient removal and harvesting strategy in urban stormwater retention ponds. *Int. Biodeter. Biodegrad.* **2016**, *112*, 80–87. [CrossRef]

41. De Stefani, G.; Tocchetto, D.; Salvato, M.; Borin, M. Performance of a floating treatment wetland for in-stream water amelioration in NE Italy. *Hydrobiologia* **2011**, *674*, 157–167. [CrossRef]

42. Mietto, A.; Borin, M.; Salvato, M.; Ronco, P.; Tadiello, N. Tech-IA floating system introduced in urban wastewater treatment plants in the Veneto region—Italy. *Water Sci. Technol.* **2013**, *68*, 1144–1150. [CrossRef]

43. Allen, R.G.; Pereira, L.S.; Raes, D.; Smith, M. *Crop Evapotranspiration. Guidelines for Computing Crop Water Requirements*; FAO Irrigation and Drainage Paper n. 56; Food and Agriculture Organisation of the United Nations: Rome, Italy, 1998.

44. Zancan, S.; Cesco, S.; Ghisi, R. Effect of UV-B radiation on iron content and distribution in maize plants. *Environ. Exp. Bot.* **2006**, *55*, 266–272. [CrossRef]

45. Göbel, P.; Dierkers, C.; Coldewey, W.G. Storm water runoff concentration matrix for urban areas. *J. Contam. Hydrol.* **2007**, *91*, 26–42. [CrossRef] [PubMed]

46. Zgheib, S.; Moilleron, R.; Chebbo, G. Priority pollutants in urban stormwater: Part 1—Case of separate storm sewers. *Water Res.* **2012**, *46*, 6683–6692. [CrossRef] [PubMed]

47. Gnecco, I.; Beretta, C.; Lanza, L.G.; La Barbera, P. Storm water pollution in the urban environment of Genoa, Italy. *Atmos. Res.* **2005**, *77*, 60–73. [CrossRef]

48. Tanner, C.C.; Headley, T.R. Components of floating emergent macrophyte treatment wetlands influencing removal of stormwater pollutants. *Ecol. Eng.* **2011**, *37*, 474–486. [CrossRef]

49. Van De Moortel, A.M.K.; Meers, E.; De Pauw, N.; Tack, F.M.G. Effects of vegetation, season and temperature on the removal of pollutants in experimental floating treatment wetlands. *Water Air Soil Pollut.* **2010**, *212*, 281–297. [CrossRef]

50. Borne, K.E.; Fassman, E.A.; Tanner, C.C. Floating treatment wetland retrofit to improve stormwater pond performance for suspended solids, copper and zinc. *Ecol. Eng.* **2013**, *54*, 173–182. [CrossRef]

51. Matthews, D.J.; Moran, B.M.; Otte, M.L. Screening the wetland plant species Alisma plantago-aquatica, Carex rostrata and Phalaris arundinacea for innate tolerance to zinc and comparison with Eriophorum angustifolium and Festuca rubra Merlin. *Environ. Pollut.* **2005**, *134*, 343–351. [CrossRef]

52. Reeves, R.D.; Baker, A.J.M. *Phytoremediation of Toxic Metals: Using Plants to Clean up the Environment*; Raskin, I., Ensley, B.D., Eds.; John Wiley and Sons Inc.: New York, NY, USA, 2000.

 *land*

Article

Urban River Recovery Inspired by Nature-Based Solutions and Biophilic Design in Albufeira, Portugal

Marie Luise Blau [1], Frieder Luz [1] and Thomas Panagopoulos [2],*

[1]   Weihenstephan-Triesdorf University of Applied Sciences, Am Hofgarten 4, 85354 Freising, Germany;
      marie-luise.blau@student.hswt.de (M.L.B); frieder.luz@hswt.de (F.L.)
[2]   Research Centre of Tourism Sustainability and Well-being, University of Algarve, 8005-139 Faro, Portugal
*   Correspondence: tpanago@ualg.pt; Tel.: +351-289-800-900

Received: 25 October 2018; Accepted: 13 November 2018; Published: 17 November 2018

Abstract: Mass urbanisation presents one of the most urgent challenges of the 21st century. The development of cities and the related increasing ground sealing are asking even more for the restoration of urban rivers, especially in the face of climate change and its consequences. This paper aims to demonstrate nature-inspired solutions in a recovery of a Southern European river that was canalised and transformed in culvert pipes. The river restoration project naturally tells the history of the city, creates a sense for the place, as well as unifying blue–green infrastructure in a symbolic way by offering areas for recreation. To improve well-being and city resilience in the long term, a regenerative sustainability approach based on biophilic design patterns was proposed. Such actions will provide greater health, social cohesion, and well-being for residents and simultaneously reduce the risks of climate change, such as heat island effect and flash floods, presenting the benefits of the transition to a regenerative economy and holistic thinking.

Keywords: built environment; urban design; regenerative design; sustainable development; river restoration; biophilic urbanism

---

1. Introduction

The global population is to reach almost 10 billion by 2050 according to revised projections [1]. Today, 54% of the world's population lives in urban areas, a proportion that is expected to increase to 66% by 2050 [2]. The underlying economic conditions and the need for growth, due to the growing population, have to include environmentally sustainable policies in order to address the problem in accordance with a healthy environment. Quality of city life and the attractiveness of cities are key parameters for success in the global competition for growth. In an effort to face the needs of our society we have to move from the idea of a circular economy towards a regenerative economy. While a circular economy is an attractive policy that aims to keep products at their highest utility through a positive developing cycle, a regenerative economy aims to create a stable and healthy system including not only green solutions but a humanistic and ecological values system that has to do with the rebirth of life itself [3].

The current system of conventional design, where we produce, consume, and create waste, placed us at the mercy of abrupt climate change and social and ecological collapse [4]. Regenerative systems, through their implicit design, do just the opposite. Regenerative economics is an economic system that works to regenerate capital assets providing goods and services that contribute to our well-being [5]. It is a principle of an ongoing self-renewal process, which built relationships and allows socio-ecological systems to constantly evolve.

Nature is regarded as vital for a fulfilled and healthy life and cities and urban planners are encouraged to advance Biophilic design: to bring more nature into the city, making it greener and richer in nature [6]. It is unanimous that the relationship between green infrastructure and the urban environment is essential to increase well-being and population health. Therefore, as mentioned by Kellert [7], landscape design that reconnects the humankind with nature is essential to provide people with opportunities to live and work in healthy places with less stress and overall greater health and well-being.

As the world population continues to urbanise and the effects of global warming are predictable, a sustainable development that encourages our city's ecosystems rather than eliminating them is the most essential. Urbanisation causes a number of problems, such as the loss of green public space for recreation, increased surface runoff, and flooding hazards and thus, negatively affects people's well-being. Urban river restoration opportunities bring nature back to the cities and help to develop climate resilient cities [8]. Currently, the European Union is aiming to be a world leader in renaturing cities and researches projects of river restoration and nature-based solutions to promote climate resilience in urban areas [9].

Nature-based solutions reduce multiple risks, as well as contribute to climate change adaptation and mitigation. River restoration using nature-based solutions can help to reduce the risk of floods and heat waves, while improving the water quality and quantity. Green infrastructure provides numerous ecosystem services that positively affect people's lives [10]. The international literature provides a new evidence-based vision enabling cities to adapt, develop and reconnect with nature [10–12].

An objective of this paper was to demonstrate nature-inspired solutions for the recovery of the river of Albufeira in Southern Portugal, which was canalised and transformed in culvert pipes. The project proposal identifies actions to move towards a Biophilic City; ways to promote nature within the built environment; ways to improve well-being and city resilience in the long term; and also proposes a regenerative sustainability approach based on biophilic design. Such actions will reduce the risks of climate change, such as heat island effect and flash floods, and present the benefits of the transition to a regenerative economy and holistic thinking.

2. Theoretical Framework of the Study

2.1. Biophilia and Biophilic Design

Biophilia is humankind's inherited biological connection with nature. According to Browning [13], the constant existence of natural elements in our historical past, human intuition, and neural science shows that connections with nature are vital to providing a healthful and dynamic life as an urban species. Most of what we regard as normal nowadays is relatively recent, such as raising food on a large scale in the last 12,000 years, the invention of the city 6000 years ago and the mass production of goods and services since the 19th century. The long passed ages of humanity's direct contact with nature explains why crackling fires fascinate us, why a garden view can enhance our creativity, why animal companionship and strolling through a park have healing effects on mental and physical health [14].

The human evolution shows mankind evolving in an adaptive response to natural and not artificial or human created forces [15]. Additionally, Kellert and Calabrese [15] mention that the human body as well as its mind and senses developed in a bio-centric and not civil engineered world. Thus, a biophilic design deals with the relationship between nature and design of the built environment, while treating our surrounding environment with respect rather than domination. The biophilic design aims to utilise green infrastructure to improve people's health and well-being [16]. Therefore, a landscape design, which reconnects humankind with nature is essential to provide people with opportunities to live and work in healthy places with less stress and overall greater health and well-being. As global warming impacts well-being and the world population continues to urbanise, a sustainable development that restores urban ecosystems rather than eliminates them is crucial.

2.2. Development of Cities and the Metamorphosis of Their Rivers

Humans used to settle down close to riverbanks, where the soil is fertile and the source of food and possibility of transport are given. Over time this settlement has transformed the natural environment into the towns, cities, and ports we see today. In that way, the increasing migration to cities leads to an unsustainable evolution, where a development that meets the needs of the present generation without compromising the ability of the future generation to meet their own needs has declined [17].

As the settlement pressure in cities increases, green spaces have to make room for human settlements, the development for trade, industry, and infrastructure. Additionally, urban rivers have been heavily reduced to enable development, to carry waste, to supply drinking water, and facilitate transport. As mentioned by Prominski et al. [18], a city's first engineering constructions were designed to regulate rivers with the purpose to protect settlements from the destructive forces of floodwater. Nowadays, most urban rivers have been straightened and transformed into channels or underground culvert pipes.

Owing to the formulation and implementation of the EU Water Framework Directive [19], which aims to achieve good conditions in all European Water courses, an increasing amount of attention was directed towards urban rivers. This was not only about flood defence as a consequence of climate change, but the opportunities offered by rivers for recreational use becoming important as places for contemplation and rehabilitation. With considerable improvements in water quality through better wastewater treatment, rainwater management, and the establishment of water-purifying plants, urban rivers are no longer shunned as stinking backwater; according to Prominski et al. [18], it is the fairest face and the first impression visitors gain of a town.

In water management terms, predictions of climate change and isolated flood and low-flow emergencies have directed attention to the necessity of adapting urban river spaces. The prognosis of longer periods of drought, more frequent heavy downpours and rising sea levels has led to the critical examination of flood protection systems and of cities' water supply and wastewater systems. The 2007 EU Flood Risk Management Directive [20] committed member states to carry out precise evaluations of the dangers posed by flooding and to draw up management plans to improve flood protection. The resulting necessary mitigation works brought the need for change to the urban environment, both above and below ground. In parallel, the EU Water Framework Directive [19], prioritised ecological objectives, such as better water quality and watercourse structure, requiring the protection, enhancement, and restoration of all surface water bodies so that cities become more sustainable in a way that allows both current and future generations to meet their needs.

2.3. Urban River Restoration for City Resilience Enhancement

River restoration is an emergent activity in many countries because of the increasing awareness of environmental degradation [21]. In Europe, urban ecosystems have been degraded as a result of human activities. The development of cities and the related increasing ground sealing further require a restoration of urban rivers, especially in the face of climate change and its consequences. Hard surfaces prevent water from naturally draining through the soil, resulting in increased floods, erosion, pollution, and decreased habitat. Urbanisation affects a river in many ways including water quality, physical structure, and the ability to support wildlife. It also influences run-off from impervious surfaces such as roads, roofs, and water quantity because of the decreased flow and reduced groundwater levels through abstraction. These stresses make rivers less resilient to the effects of climate change, which would further have a negative effect on the cities' climate [22].

The simplest method to improve rivers is to return flows to a more natural state. Implementing a green-blue infrastructure, including sustainable urban drainage and green roofs, is a long-term approach to managing surface and groundwater by reducing the rate and volume of run-off [23]. In that way, river restoration directly improves the habitat, reduces the risk of flash floods, improves groundwater management, and decreases the urban heat island effect.

However, the consequences of catastrophic floods of the recent years, the need for more public green space for adaptation to climate change risks and the perception that we have to restore part of the damaged urban ecosystems have increasingly been called into question [24]. Urban regeneration is an opportunity to revitalise rivers and offers a variety of ecological, social, and economic benefits including an improved flood management using more natural processes, a reduced likelihood of negative impacts caused by climate change through increased ecosystem resilience, as well as a reconnection of people to the natural environment through better access to recreation and improved well-being [25].

The Cheonggyecheon river restoration project was an ambitious urban regeneration initiative that transformed the urban space of Seoul, Korea [26]. The river that had been buried under a highway in 1967 was recovered in 2002 by decommissioning the highway, excavating the new river channel and undertaking works for a linear park river corridor that decreased flood risk and created recreation opportunities [27]. The Madrid Rio Project was another urban regeneration project that transformed one of the most degraded and neglected zones of the city and became one of the most beautiful cultural areas. It connected green zones and historic gardens and recovered the use of the river [28].

2.4. Regenerative Design for the Degraded Urban Ecosystem

Restorative sustainability employs strategies in the city process of design that produce a positive impact on the natural environment, society, and human well-being, restoring socioecological systems to a healthy state [29]. Many European cities are frontrunners in the transition towards a low carbon, resource efficient, and green economy. Meanwhile, it is becoming clear that the built environment must go beyond this. It must have a net positive environmental benefit for the living world [30]. Creativity and innovation will build a new economic system to address urban growth problems in an entirely different direction. Regenerative design is an effort to build a green economy that restores the relationship between nature and people.

Key challenges for sustainable cities are to provide solutions to significantly increase cities' resource efficiency through actions addressing mobility, climate change, and environmental quality [31]. Such actions should bring profound economic, social and environmental impacts, resulting in a better quality of life (including health and social cohesion), jobs, and growth [32]. A range of design solutions can be adopted to improve aspects of urban design by reducing the heat island and other climate change effects [33]. The role of plants in mitigating and adapting the urban environment to climate change is mentioned in many studies about regenerative sustainability and resilient cities [34].

The regenerative design follows some principles by designing in an adaptable way without wasting extra materials, to restore, to reuse, to remanufacture, and finally to recycle. By using environmentally friendly adapted and reconstructed materials, we create a more regenerative built environment. Figure 1 presents the stages from conventional to regenerative design and consequently from conventional to regenerative economy. The main idea is to use holistic thinking instead of reductionist thinking, to move from technical systems development to living systems development, and from degenerating design to regenerative design to build into the landscape using patterns that connect nature and humans in a coevolving relationship [35].

Figure 1. The stages of development, from conventional to regenerative, showing that regeneration goes far beyond sustainability (adapted from References [36,37]).

The built environment is part of the climate change problem, thus, sustainability in green infrastructure has to be recognised as an adequate measure for current and future urban design. In this respect, regenerative urban systems have to address climate change; limited resources and social divide aiming not only to reduce but also to create positive impacts. Urban green infrastructure has to do with the improvement in energy, water, and carbon resources consumption, which will have positive effects on the place, health, and education. Thermal comfort and energy efficiency, net positive water, recycled or upcycled materials, a safe environment, good relationship with the place, positive feelings, and environmental justice are some of the positive effects of regenerative economy systems [38,39].

3. Materials and Methods

3.1. The Study Area

Albufeira is a small city in the southernmost Portuguese region of the Algarve. The municipality population in 2011 was 40,828, in an area of 140 km². A sea and sun tourist destination, Albufeira expands to more than 300,000 residents during summer. The climate is Mediterranean with hot and dry summers, an average temperature of 17.7 °C, 500 mm of precipitation (the highest month is January with 84 mm) and more than 300 sunny days. The Albufeira River crosses the old town and ends in Pescadores beach. This river creates frequent events of flash flooding because it was canalised and put underground in sewage pipes close to the old town, which according to studies of the regional hydrographic administration, were insufficient (>50 m³/s instead of 86–130 m³/s of expected runoff).

The river of Albufeira flows its natural course in a soft meander between low hills (Figure 2) ending in a small lagoon next of the beach in the Atlantic Ocean. Earlier the city was called Al-uhera by the Arabs when it was part of the Arab empire, which means that the coastal lagoon historically was a defining factor for the city. Before the tourism boom, Albufeira was a small fishing village with strong connections to the ocean and a fishery economy. Albufeira was a naturally built retention basin in the middle of the city, acting furthermore as a city harbour where the ecological water cycle managed heavy rain events by itself.

As urbanisation increased in Albufeira due to high demand from tourism, the lagoon was replaced by the new city centre, which is completely impermeable to water and does not reflect the sense of the place, supporting flash floods rather than preventing them. Rebuilding the natural river ecosystem running through the city will give Albufeira its individual sense back as well as protecting the city from flood hazards due to a biophilic urban design.

Figure 2. A representation of the medieval times city of Albufeira and a recent photo from the flash flood (obtained from the RTP TV channel during the news of 1 November 2015), showing that the buried river found its natural way to the Atlantic Ocean.

3.2. Analysing the Study Area by the Science of Strolling

The new proposal for Albufeira River recovery has to meet the main objectives, such as the creating of a sense of place, a biophilic alternative facing flood protection, and improving well-being, as well as socio-economic advantages of nature-based solutions.

The study area was analysed by the science of strolling, which is a method in the field of aesthetics and cultural studies with the aim of becoming conscious of the conditions of perception of the environment and enhancement of environmental perception itself. According to Burckhardt [40], the critical analysis of contemporary planning practices, through the perspective of a walker at the study area, can be characterised by four different zones each marked by their surrounding elements with conflicts as well as potentials arising out of the site-specific atmosphere of the surroundings as shown in Table 1. Figure 3 illustrates the four zones of the study area following the science of strolling.

Table 1. The conflicts and potentials of the study area.

Zones	Conflict	Potential
Zone 1: Backyard silence	Unshaded plain park area without identifying the character, no integration of the river and the bordering camping area	Extending the atmosphere of the densely planted camping area with its community gardens by integrating the river as a connecting element and thus, creating a cooling comfort as well as a backyard feeling for recreation
Zone 2: Sub-urban laughter	Lined up landscape elements (residential—areapark—road) without any interaction and the disappearing of the river from view	Making the residential area interact with the park due to the creation of a joyful public area with the river integration as the main attraction as well as providing a sound and visual barrier for the bordering road linking both sides by a pedestrian underpass
Zone 3: Nature chirping	Busy road separating the municipal green belt and negatively influencing the recreational nature experience the character of the area	Connecting the green belt by implementing a green bridge for animal migration as well as extending the forest atmosphere of the surrounding mountains by the creation of urban woods and wetlands, providing a retention basin tangible in all states of flood levels
Zone 4: Urban sounds	Heat island effect strengthened by the absence of the river and a climate adapted infrastructure, as well as the lack of a historical sense of the place	Recreating the sense of place as well as linking public urban spaces by following the symbolic waterline acting at the same time as a part of the city's blue and green infrastructure

Figure 3. The four zones of the study area following the science of strolling: Area 1-backyard silence (**A**); Area 2—sub-urban laughter (**B**); Area 3—nature chirping (**C**); Area 4—urban sounds (**D**).

The strolling process is an alternative to the reductionist thinking and technocratic centrally planned economy. It is based on traditional methods in cultural studies as well as experimental practices like taking reflective walks and aesthetic interventions [40].

4. Results

4.1. The Concept of the Project

The concept of the Albufeira project for urban river recovery was inspired by nature-based solutions and biophilic design. It was inspired by previous research on biophilic design, the development of cities and their rivers as well as nature-based solutions for city resilience enhancement and the analysis of the study area using the strolling method. The river recovery was conceived as a continuous green corridor along the restored river evolving all the way down from the camping area north of Albufeira until the river mouth in the Atlantic Ocean.

Making the city resilient to climate change and its consequences, linking the bordering residential areas, activating city dweller and visitor interest to the history of the place, and the provisioning of areas for recreation were positive byproducts of the river restoration using high standards in the socioeconomic competition of the touristic hotspots in the Algarve region. Turning the fear of the river's destructive force into an atmospheric enjoyment, tangible by strolling along the symbolic waterline, strengthens the connection of humankind and the river ecosystem; improving quality of life and environmental stewardship. The outcome was the creation of a new linear park for Albufeira that is ecologically sound, aesthetically satisfying, and economically rewarding. The project connects residents and tourists with nature and the spirit of the place providing them with outdoor recreation opportunities in a multifunctional space that enhances city resilience (Figure 4).

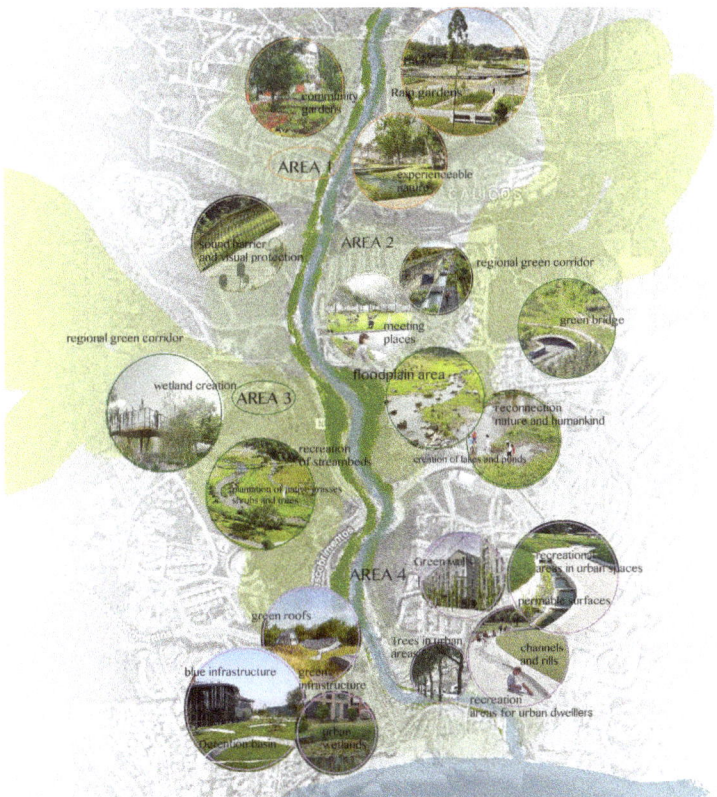

Figure 4. The concept of Albufeira's green corridor linear park.

4.2. Recreation for City Dwellers and Visitors Due to a Biophilic Design

As already asserted in 1877 by John Muir, the people receive from nature far more than what they seek. Biophilic design reduces stress, improves cognitive function and creativity, improves our well-being, and provides healing for both city dwellers and visitors [13]. Adapted from Browning et al. [13], Table 2 illustrates the patterns of biophilic design and its functions in supporting stress reduction, cognitive performance, emotion, and mood enhancement as well as the human body in general. These patterns were used to provide recreation areas by walking through the new environment evolved by the restoration of the Albufeira River and the enhancement of its accompanying linear park.

Strolling through area 1, people experience the intimate, familiar, and calm backyard atmosphere given by the bordering camping area, which is enhanced by the implementation of fruit meadows, raingardens, and constructed wetlands containing water purifying plants that provide a better water quality along the reintegrated river (Figure 5a). Providing sports areas and meeting points along a curved pathway creates a slow and joyful movement through space inviting strollers to explore the environment. Due to the connection with nature and its natural river ecosystem, the presence of water as well as biomorphic forms of the pathway, the meeting areas and the shape of the river, this area enjoys all the advantages for recreation as shown in Table 2.

Table 2. The patterns of a biophilic design and their functions in supporting stress reduction, cognitive performance, emotion, mood and preference (adapted from Reference [13]).

Patterns	Stress Reduction	Cognitive Performance	Emotion, Mood and Preference
Visual connection with nature. A view to elements of nature, living systems, and natural processes	Lowered blood pressure and heart rate	Improved mental engagement/attentiveness	Positively impacted attitude and overall happiness
Non-visual connection with nature. Auditory, haptic, olfactory, or gustatory stimuli that create a deliberate reference to nature	Reduced systolic blood pressure and stress hormones	Positively impacted on cognitive performance	Perceived improvements in mental health and tranquillity
Thermal and airflow variability. Subtle changes in air temperature, relative humidity, airflow that mimic natural environments	Positively impacted comfort, well-being and productivity	Positively impacted concentration	Improved perception of temporal and spatial pleasure
Presence of water. A condition that enhances the experience of a place through seeing, hearing or touching of water	Reduced stress, increased feelings of tranquillity,	Improved concentration and memory restoration	Observed preferences and positive emotional responses
Connection with natural systems. Awareness of natural processes, seasonal changes, characteristic of a healthy ecosystem		Enhanced perception and psychological responsiveness	Enhanced positive health responses; shifted perception of the environment
Biomorphic forms and patterns. Symbolic references to contoured, patterned, textured or numerical arrangements that persist in nature	Lower heart rate and blood pressure		Observed view preference
Material connection with nature. Material and elements from nature that create a sense of place	Decreased diastolic blood pressure	Improved creative performance	Improved comfort
Complexity and order. Rich sensory information that adheres to a spatial hierarchy similar to those encountered in nature	Positive perceptual and physiological stress responses	Improved concentration, attention	Observed view preference
Prospect. An unimpeded view over a distance for surveillance and planning	Reduced stress	Reduced fatigue	Improved comfort and perceived safety
Refuge. A place for withdrawal from environmental conditions in which the individual is protected from behind and overhead	Reduced stress	Perception of safety	Improved comfort and perceived safety
Mystery. The promise of more information achieved through partially obscured views that tempt to travel deeper into the environment		Improved creative performance	Induced strong pleasure response

By enhancing the connection of the residential area and the bordering park, bringing the river above ground, implementing playgrounds and thus, creating a sub-urban atmosphere in area 2, one can be active and interact with neighbours and city visitors and at the same time find relaxation enjoying the green lawns provided with seating steps facing the river. Well-being and greater health can be experienced in a friendly atmosphere by creating complexity and order due to plant selection variety and different pedestrian flows created by the movement through the space (fast or slow) and the path hierarchy (narrow and wide) (Figure 5b). A mysterious and peaceful atmosphere can be experienced in the nature chirping atmosphere. Urban forests and wetlands pick up the surrounding atmosphere of the densely planted mountains making visitors explore the landscape in all flood levels due to a pedestrian footbridge reaching into the retention area where water can be stored in case of occurring flash floods. As the Albufeira beach does not allow city visitors to bring their dogs, a dog park was built in this area offering dog lagoons containing naturally cleaned water by the close-by urban wetlands. The promise of more information achieved through partially obscured views due to the implementation of the forests and the hidden movement through the space, entice the individual to travel deeper into the environment and gain recreation before entering the rushing city area.

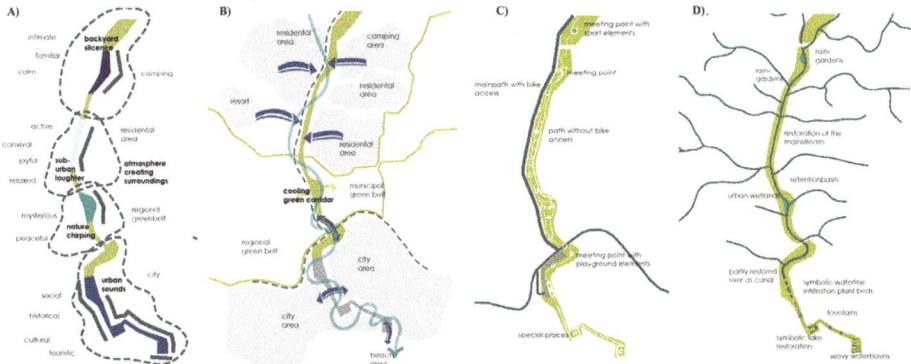

Figure 5. The representation of the atmosphere (**A**); green inking corridor (**B**); public spaces and path network (**C**); river restoration and the presence of water (**D**) in the four zones as defined by the strolling method.

One reaches zone 4 by walking through a pedestrian underpass designed as a wide and light tunnel where humankind experiences the non-visual connection with audible nature by the played sound of the water to avoid fear while passing it (Figure 5c). By experiencing the urban sounds, one finds out that the shape of the river changes, as the water is now running through a channel. The further humankind gets to the city centre, the more the river turns into a symbolic waterline represented as infiltration and evaporation beds, fountains, and water basins leading the way to the ocean where the river once used to run its natural course. As the functions of urban areas are spread wide, this zone can be roughly subdivided into three parts starting with the new mobility centre offering a bike and car park, as well as a local bus station and a green-roofed parking deck, letting the street end at this point. The multi-functional use of this area offers space for events as well as green fingers reaching into the city area providing a cooling comfort. By reducing the heat island effect, a blue and green infrastructure reaches all the way from the mobility centre along the symbolic waterline to the beach area where the river once found its mouth in the Atlantic Ocean (Figure 5d).

4.3. Creating a Sense of Place

Designed as a shaded cooling area, water basins are wave-shaped, symbolising the strong connection to the ocean Albufeira had as a small fishing village. As the conflict-potential is the highest in the urban sound area, this zone containing the city centre of Albufeira requires a more detailed proposal linking the following three dimensions: atmosphere, function, and space (Figure 6).

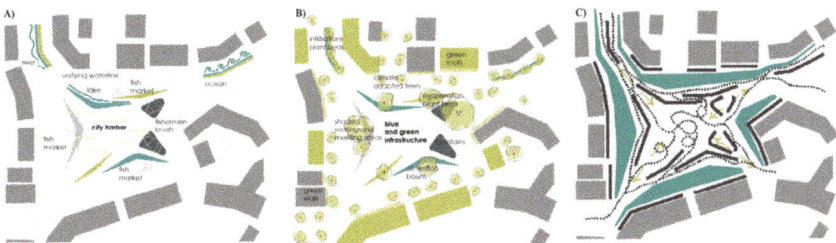

Figure 6. The three dimensions (atmosphere (**A**), function (**B**), space (**C**)), linked in the city centre creating a sense of place and bounding elements to tell the story of Albufeira's history and reminding people of the name of the city in Arabic, Al uhera (the lagoon by the sea), in a symbolic way.

As described by Weidinger [41], what identifies a place is the character of the location made out of its space (spatial borders, structure, visual axes), function (cooling, entertaining, recreating) as well as its atmosphere. Known as a small fishing village, the fishing economy of Albufeira was always of high importance for the city. The naturally built lake in the middle of the old town centre was acting as a small harbour, where fishermen used to sell their day's catch. As shown in Figure 6, the space bounding elements tell the story of Albufeira's history, creating a sense for the place as well as unifying all bodies of surface water once represented in the city in a symbolic way. Turning people's fear of water into a "safe harbour" due to the pleasant presence of water creates a sense of the place offering areas for recreation.

The city centre should be multifunctional and enhance resilience to climate change and its consequences. In that aspect, creating a blue and green infrastructure for the city, as illustrated in Figure 6, reducing the heat island effect as well as providing temporary storage for occurring heavy rain events was the main objective of this area, together with the entertaining and recreating functions.

In Figure 6 the surrounding buildings of the city centre can be seen (mainly restaurants and tourism-related shops), which characterise this urban public space, and narrow streets that connected the inside function of the building with its outside area, interacting with the new design and building functional borders. Those borders, furthermore, create visual axes, a movement through the space as well as a public space network providing easy orientation for visitors (Figure 7, bottom-left). The new design for Albufeira's city centre brings the three dimensions together, as well as creating recreation areas. Moreover, due to a range of different biophilic design patterns, such as the presence of water and a comfortable thermal and airflow variability, a place that will provide recreation to residents and visitors of the city and be remembered can be created.

Regenerative design accepts and promotes 'place' as the primary starting point for design and connecting people back to the spirit of place in a way that they vitalised by it and become intrinsically motivated to care for it [40]. Genius Loci is the spirit, character, or atmosphere of a place or an entire town [42]. It is made out of its history and understanding of people creating an identity for the place [43]. Knowing a city's history is necessary to understand their contemporary shape and thus, realising why extreme events like flood hazards arrive from time to time.

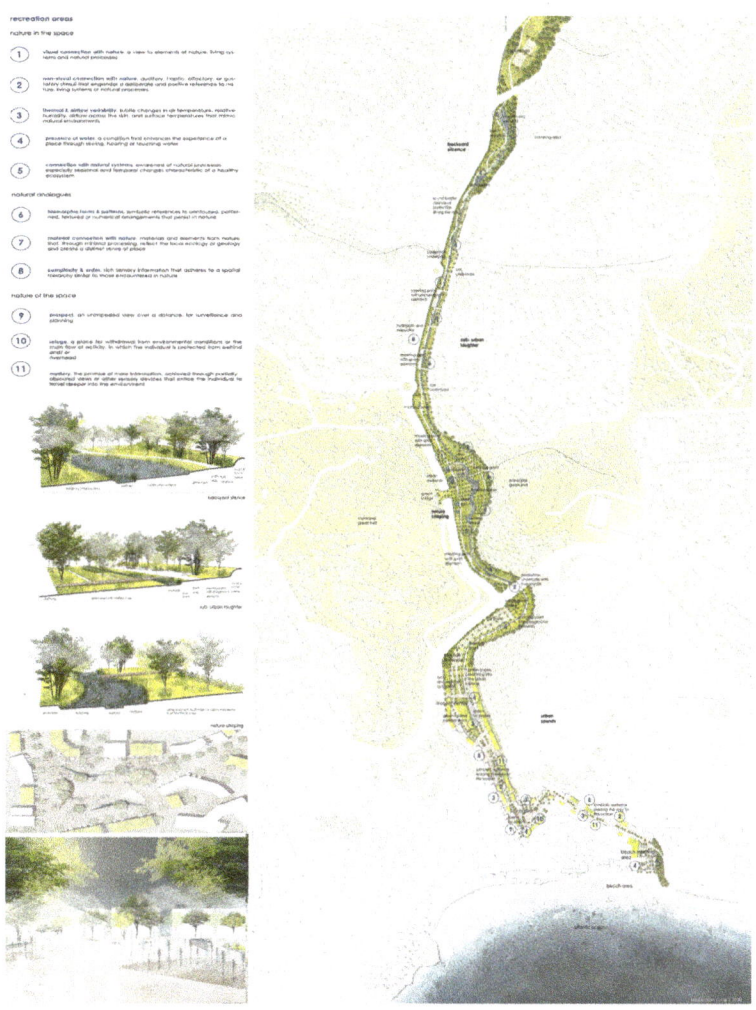

Figure 7. The urban river recovery inspired by nature-based solutions and biophilic design in Albufeira, Portugal. Perspectives from the four zones derived from the scrolling method and the city centre symbolic lagoon.

4.4. Nature-Based Solutions

Nature-based solutions to increase the well-being and city resilience to climate change were also applied at selected locations in the new linear park of Albufeira (Figure 7). At the northern part of zone 1 (backyard silence), a constructed wetland will purify the water before it enters the linear park and the following raingardens and meadows will contribute to decreasing the runoff from the nearby avenue. Trees and shrubs along the avenue will offer privacy screening, acting as a sound and visual barrier (zone 2). Permeable paving options will be used for the paths along the park, while glowing stones will absorb solar energy during the day and release it at night at the cycle path, as in the Van Gogh-Roosegaarde route in Eindhoven [44]. For soil permeability and runoff reduction in the parking areas, permeable pavements may be used for a grass grid so the water could infiltrate and decrease the risk of flooding [45]. Retention basins and urban wetlands are proposed for zone 3 (urban chirping).

The region of the Algarve is a privileged place to reuse the solar energy, for that reason, solar benches and solar lights are some examples of the utilisation of sustainability techniques in the park. For zone 4 (urban sounds), green walls are proposed, which may decrease the visual and noise impact, and green rooftops on buildings may provide multiple ecosystem services such as thermal comfort, energy efficiency, retention of stormwater, and the provision of habitats. Nature-based solutions may bring more diverse natural features and processes into the urban area through locally adapted and resource efficient interventions [46].

Regenerative projects aim to reconnect the city and nature, improving quality of life and environmental stewardship [47]. A recycled landscape, which provides different multipurpose uses, is attractive and viable [48]. A good example of a landscape reclamation project was Park-Trançâo, realised in Portugal at the Lisbon River [49]. This derelict industrial area was ecologically degraded, with high soil, water, and air contamination problems. The restoration of the 'place' offered a good public space for the well-being of citizens and also a healthy relationship between city and river.

One of the regenerative economy targets is to minimise the use of nonrenewable natural resources [50]. Therefore, in order to ameliorate the quality of city life and make cities attractive, flexible building materials, which last and adapt, and upcycling strategies are solutions for sustainable urban design [51]. One of the most important resources in water and regenerative design should recycle and purify all urban water by introducing new innovative technologies. Cities should increase green infrastructure and decrease sealed areas preventing the functioning of the water cycle [46]. Energy is another sector where long-term potentials and regenerative actions are required for sustainable development [52]. Solar and wind renewable energy can be used to decrease the CO_2 emissions and prevent the consequences of global warming from greenhouse gas emissions, meeting the emissions goal. Moreover, increasing urban green infrastructure has efficient results in reducing carbon emissions and many benefits for everyone, especially to economically vulnerable groups of people [53].

5. Conclusions and Recommendations

This paper presented a case of biophilic urbanism using the project of the Albufeira city river restoration. The project was inspired by nature-based solutions and by implementing a green and blue rather than a grey infrastructure to reduce the risks of climate change, such as heat island effect and flash floods, while simultaneously providing greater health and well-being for urban residents and visitors. The new urban design for Albufeira reflects humanity's innate need for nature and meets the needs of the following generations in a long-term sustainable solution that provides retention basins for water storage after torrential rains as well as a green and blue infrastructure, allowing the water to run its natural cycle. Due to the construction of wetlands and raingardens, the water quality will be improved and will enter the ocean clean. Adjacent to the supply of ecosystem services, a range of environmental, socio-cultural, ecological, and economic advantages were a result of the new design. The cooling green corridor brings fresh air to the city and thus, reduces the heat island effect making the city more energy efficient. City-cooling places all along the symbolic waterline, close to public spaces in the urban sounds area, create high-quality places for city residents and visitors. By establishing the blue-green infrastructure, the city will be cooled down, providing a great advantage in the socio-economic competition of the touristic hot spots in the region.

Another great advantage of the project was the creation of a historical sense of individuality, bringing back the identity of a place that will be memorised and associated with Albufeira and its fishing history. The project created a sense of place by recovering the buried river, reconnecting citizens with nature and the history of the place. Strolling along the symbolic waterline, the connection of people with the river ecosystem was strengthened and the fear of the river's destructive force tangibly turned into an atmosphere for enjoyment. The outcome was the creation of a linear park for Albufeira that was ecologically sound, aesthetically satisfying, economically rewarding, and where the relationship between people and nature was of love and respect rather than domination.

A civil engineering alternative proposal that was examined by the municipality of Albufeira pretended to solve the frequently occurring flash flood problems by constructing a tunnel to increase the drainage capacity and circumvent the city. This grey infrastructure solution, which might be technically correct, rapid, and with minimum environmental impact, does not bring any other socioeconomic advantage for the town and its population. The green and blue infrastructure provides multiple socio-cultural benefits and ecosystem services. Furthermore, it might be less cost-intense and bring long-term benefits that will enhance city resilience to climate change and improve human health and well-being. The alternative project coming from civil engineering reductionist thinking has a short-term focus at the expense of long-term effects, while holistic thinking approaches river recovery as part of the long-term city planning for climate change adaptation in the implementation of nature-based solutions in a biophilic design project. This was seen as an opportunity to improve well-being and increase city resilience instead of providing solutions for flood risk.

Currently, there is no available data for the comparative assessment of the civil engineering and biophilic proposals and a quantitative framework of key performance indicators will be a task for future research. In order to provide evidence of the advantages of the solutions compared to alternative options needs monitoring for a long period since the benefits and services provided by the urban green infrastructure do not take place immediately after the completion of the project. Additionally we need to develop cost-benefit methodologies to assess the impact of the deployed solutions in as quantifiable a way as possible and considering all benefits (such as carbon sequestration, mitigation of heat island effects, natural cooling and heating, recreation, mitigation of soil sealing effects, enhanced soil, flood prevention, enhancement of biodiversity and natural capital, human well-being and health, reduction of noise and air pollution, improvement of water quality, and others [46]).

Taking account of scarce financial resources available to local governments and the lack of data to support the long-term benefits, it is necessary to find new approaches for financing this kind of project. Taking into consideration the uncertainties on cost, risks, and benefits, governments should promote biophilic urbanism and nature-based solutions by regulations and lower taxation, so as to make them attractive to citizens and companies, boosting the whole economy. More research is needed on biophilic urban planning approaches for different size cities, located in various climatic zones, with diverse ethnic compositions, income, quality of urban life etc.

Author Contributions: Conceptualization, M.L.B. and T.P.; Methodology, M.L.B. and T.P; Formal Analysis, M.L.B.; Investigation, M.L.B..; Writing—Original Draft Preparation, M.L.B. and T.P.; Writing—Review & Editing, T.P. and F.L.; Supervision, T.P. and F.L.

Funding: This research was funded by the Foundation for Science and Technology grant number PTDC/GES-URB/31928/2017 and co-funded by the Erasmus+ programme of the European Union.

Acknowledgments: The authors would like to acknowledge the range of inputs and support received from the Albufeira Municipality. The BIODES programme was funded the Foundation for Science and Technology through project PTDC/GES-URB/31928/2017 "Improving life in a changing urban environment through Biophilic Design". Marie Luise Blau and Frieder Luz acknowledge the Erasmus+ programme of the European Union.

References

1. United Nations. The World Population Prospects: The 2017 Revision. Available online: https://esa.un.org/unpd/wpp/ (accessed on 12 December 2017).
2. UN. 2014. Available online: http://www.un.org/en/development/desa/news/population/world-urbanization-prospects-2014.html (accessed on 12 December 2017).
3. Lyle, J.T. *Regenerative Design for Sustainable Development*; Wiley: New York, NY, USA, 1994.
4. Hens, L. The challenge of the sustainable city. *Environ. Dev. Sustain.* **2010**, *12*, 875–876. [CrossRef]

5.   Kopeva, D.; Panagopoulos, T.; Villa, J.K.; Stasiskiene, Z.; Shterev, N.; Baltov, M. Circular economy. In *Sustainability, Restorative to Regenerativ*; Haselsteiner, E., Apró, D., Kopeva, D., Eds.; ERAC Research: Bolzano, Italy, 2018; pp. 86–95. Available online: http://www.eurestore.eu/wp-content/uploads/2018/04/Sustainability-Restorative-to-Regenerative.pdf (accessed on 7 October 2018).

6.   Beatley, T. *Biophilic Cities: Integrating Nature into Urban Design and Planning*; Island Press: Washington, DC, USA, 2011.

7.   Kellert, S. Biophilic urbanism: The potential to transform. *Smart Sustain. Built Environ.* **2016**, *5*, 4–8. [CrossRef]

8.   Giannakis, E.; Bruggeman, A.; Poulou, D.; Zoumides, C.; Eliades, M. Linear parks along urban rivers: Perceptions of thermal comfort and climate change adaptation in Cyprus. *Sustainability* **2016**, *8*, 1023. [CrossRef]

9.   Raymond, C.M.; Berry, P.; Breil, M.; Nita, M.R.; Kabisch, N.; de Bel, M.; Enzi, V.; Frantzeskaki, N.; Geneletti, D.; Cardinaletti, M.; et al. *An Impact Evaluation Framework to Support Planning and Evaluation of Nature-based Solutions Projects*; Centre for Ecology & Hydrology: Wallingford, UK, 2017.

10.   Lovell, S.T.; Taylor, J.R. Supplying urban ecosystem services through multifunctional green infrastructure in the United States. *Landsc. Ecol.* **2013**, *28*, 1447–1463. [CrossRef]

11.   Tzoulas, K.; Korpela, K.; Venn, S.; Yli-Pelkonen, V.; Kaźmierczak, A.; Niemela, J.; James, P. Promoting ecosystem and human health in urban areas using green infrastructure: A literature review. *Landsc. Urban Plan.* **2007**, *81*, 167–178. [CrossRef]

12.   Connop, S.; Vandergert, P.; Eisenberg, B.; Collier, M.J.; Nash, C.; Clough, J.; Newport, D. Renaturing cities using a regionally-focused biodiversity-led multifunctional benefits approach to urban green infrastructure. *Environ. Sci. Policy* **2016**, *62*, 99–111. [CrossRef]

13.   Browning, W.D.; Ryan, C.O.; Clancy, J.O. *14 Patterns of Biophilic Design*; Terrapin Bright Green LLC: New York, NY, USA, 2014.

14.   Kellert, S. *Building for Life: Understanding and Designing the Human-Nature Connection*; Island Press: Washington, DC, USA, 2005.

15.   Kellert, S.; Calabrese, E. The Practice of Biophilic Design. Available online: https://www.biophilic-design.com (accessed on 15 May 2018).

16.   Beatley, T.; Newman, P. Biophilic cities are sustainable, resilient cities. *Sustainability* **2013**, *5*, 3328–3345. [CrossRef]

17.   World Commission on Environment and Development. *Our Common Future*; Oxford University Press: Oxford, UK, 1987.

18.   Prominski, M.; Stokman, A.; Zeller, S.; Stimberg, D.; Voermanek, H.; Bajc, K. *River Space Design. Planning Strategies, Methods and Projects for Urban River*; Birkhauser: Basel, Switzerland, 2012.

19.   The EU Water Framework Directive. Available online: http://doi.org/10.2779/75229 (accessed on 2 April 2018).

20.   European Commission. The EU Flood Risk Management Directive. Available online: http://ec.europa.eu/environment/water/flood_risk/index.htm (accessed on 2 April 2018).

21.   Clewell, A.F.; Aronson, J. Motivations for the restoration of ecosystems. *Conserv. Biol.* **2006**, *20*, 420–428. [CrossRef] [PubMed]

22.   European Center for River Restoration. Urban River Restoration. Available online: http://www.ecrr.org/RiverRestoration/UrbanRiverRestoration/tabid/3177/Default.aspx (accessed on 17 April 2018).

23.   Del Tánago, M.G.; De Jalón, D.G.; Román, M. River restoration in Spain: Theoretical and practical approach in the context of the European water framework directive. *Environ. Manag.* **2012**, *50*, 123. [CrossRef] [PubMed]

24.   Berte, E.; Panagopoulos, T. Enhancing city resilience to climate change by means of ecosystem services improvement: A SWOT analysis for the city of Faro, Portugal. *Int. J. Urban Sustain. Dev.* **2014**, *6*, 241–253. [CrossRef]

25.   Loures, L.; Panagopoulos, T.; Nunes, J.; Viegas, A. Learning from practice: Using case-study research towards post-industrial landscape redevelopment theory. *WIT Trans. Ecol. Environ.* **2012**, *167*, 23–32. [CrossRef]

26.   Lee, J.Y.; Anderson, C.D. The restored Cheonggyecheon and the quality of life in Seoul. *J. Urban Technol.* **2013**, *20*, 3–22. [CrossRef]

27. Temperton, V.M.; Higgs, E.; Choi, Y.D.; Allen, E.; Lamb, D.; Lee, C.S.; Harris, J.; Hobbs, R.J.; Zedler, J.B. Flexible and adaptable restoration: An example from South Korea. *Restor. Ecol.* **2014**, *22*, 271–278. [CrossRef]

28. Martí, P.; Mayor, C.G.; Melgarejo, A. Waterfront landscapes in Spanish cities: Regeneration and urban transformations. *WIT Trans. Built Environ.* **2018**, *179*, 45–56. [CrossRef]

29. Robinson, J.; Cole, R.J. Theoretical underpinnings of regenerative sustainability. *Build. Res. Inf.* **2015**, *43*, 133–143. [CrossRef]

30. Jenkin, S.; Zari, P.M. *Rethinking Our Built Environments: Towards a Sustainable Future*; Ministry for the Environment: Wellington, New Zealand, 2009.

31. Bulkeley, H.; Betsill, M. Rethinking sustainable cities: Multilevel governance and the 'urban' politics of climate change. *Environ. Politics* **2005**, *14*, 42–63. [CrossRef]

32. Neirotti, P.; De Marco, A.; Cagliano, A.C.; Mangano, G.; Scorrano, F. Current trends in Smart City initiatives: Some stylized facts. *Cities* **2014**, *38*, 25–36. [CrossRef]

33. Santamouris, M. *Energy and Climate in the Urban Built Environment*; James and James: London, UK, 2001.

34. Karyono, T.H.; Vale, R.; Vale, B. *Sustainable Building and Built Environments to Mitigate Climate Change in the Tropics*; Tanri Abeng University: Jakarta, Indonesia, 2015.

35. Beatley, T. *Handbook of Biophilic City Planning & Design*; Island Press: Washington, DC, USA, 2016.

36. Van der Ryn, S.; Cowan, S. *Ecological Design*; Island Press: Washington, DC, USA, 1996.

37. Reed, B. Shifting from 'sustainability' to regeneration. *Build. Res. Inf.* **2007**, *35*, 674–680. [CrossRef]

38. Cole, R.J. Regenerative design and development: Current theory and practice. *Build. Res. Inf.* **2012**, *40*, 1–6. [CrossRef]

39. Pedersen, Z.M. Ecosystem services analysis for the design of regenerative built environment. *Build. Res. Inf.* **2012**, *40*, 54–64.

40. Burckhardt, L. *Warum ist Landschaft Schön?* Die Spaziergangswissenschaft Verlag: Kassel, Germany, 2006.

41. Weidiniger, J. *Atmosphären Entwerfen*; TU Berlin University: Berlin, Germany, 2016.

42. Panagopoulos, T. Linking forestry, sustainability and aesthetics. *Ecol. Econ.* **2009**, *68*, 2485–2489. [CrossRef]

43. Loures, L.; Panagopoulos, T.; Burley, J.B. Assessing user preferences on Brownfield regeneration. The case of Arade river waterfront, South Portugal. *Environ. Plan. B Plan. Des.* **2016**, *43*, 871–892. [CrossRef]

44. Van Gogh-Roosegaarde Cycle Path. Available online: https://www.heijmans.nl/en/projects/van-gogh-roosegaarde-cycle-path/ (accessed on 12 May 2016).

45. Sousa Silva, C.; Lackóová, L.; Panagopoulos, T. Applying sustainability techniques in eco-industrial parks. *WIT Trans. Ecol. Environ.* **2017**, *210*, 135–145. [CrossRef]

46. Nature-Based Solutions and Re-Naturing Cities. Available online: http://ec.europa.eu/transparency/regexpert/index.cfm?do=groupDetail.groupDetail&groupID=3164 (accessed on 19 May 2017).

47. Revell, G.; Anda, M. Sustainable Urban Biophilia: The Case of Greenskins for Urban Density. *Sustainability* **2014**, *6*, 5423–5438. [CrossRef]

48. Loures, L.; Panagopoulos, T. Sustainable reclamation of industrial areas in urban landscapes. *WIT Trans. Ecol. Environ.* **2007**, *102*, 791–800. [CrossRef]

49. Loures, L.; Panagopoulos, T. From derelict industrial areas towards multifunctional landscapes and urban renaissance. *WSEAS Trans. Environ. Dev.* **2007**, *3*, 181–188.

50. Brown, M.; Haselsteiner, E.; Apró, D.; Kopeva, D.; Luca, E.; Pulkkinen, K.; Vula Rizvanolli, B. *Sustainability, Restorative to Regenerative*; ERAC Research: Bolzano, Italy, 2018.

51. Herman, K.; Sbarcea, M.; Panagopoulos, T. Creating green space sustainability through low-budget and upcycling strategies. *Sustainability* **2018**, *10*, 1857. [CrossRef]

52. Chen, W.Y. The role of urban green infrastructure in offsetting carbon emissions in 35 major Chinese cities: A nationwide estimate. *Cities* **2015**, *44*, 112–120. [CrossRef]

53. Panagopoulos, T.; González Duque, J.A.; Bostenaru Dan, M. Urban planning with respect to environmental quality and human well-being. *Environ. Pollut.* **2016**, *208*, 137–144. [CrossRef] [PubMed]

 *land*

Article

Visitor Satisfaction with a Public Green Infrastructure and Urban Nature Space in Perth, Western Australia

Jackie Parker [1] and Greg D. Simpson [2,*]

1   School of Design and Built Environment, Curtin University, Perth, WA 6102, Australia;
    17966131@student.curtin.edu.au
2   College of Science, Health, Engineering and Education—Environmental and Conservation Sciences,
    Murdoch University, Perth, WA 6150, Australia
*   Correspondence: G.Simpson@murdoch.edu.au

Received: 31 October 2018; Accepted: 12 December 2018; Published: 17 December 2018

Abstract: The widely applied Importance-Performance Analysis (IPA) provides relatively simple and straightforward techniques to assess how well the attributes of a good or service perform in meeting the expectations of consumers, clients, users, and visitors. Surprisingly, IPA has rarely been applied to inform the management of urban public green infrastructure (PGI) or urban nature (UN) spaces. This case study explores the visitor satisfaction levels of people using a PGI space that incorporates UN, close to the central business district of Perth, Western Australia. With diminishing opportunities to acquire new PGI spaces within ever more densely populated urban centers, understanding, efficiently managing, and continuously improving existing spaces is crucial to accessing the benefits and services that PGI and UN provide for humankind. An intercept survey conducted within the Lake Claremont PGI space utilized a self-report questionnaire to gather qualitative and quantitative data (n = 393). This case study demonstrates how the IPA tool can assist urban planners and land managers to collect information about the attributes of quality PGI and UN spaces to monitor levels of service, to increase overall efficiency of site management, to inform future management decisions, and to optimize the allocation of scarce resources. The satisfaction of PGI users was analyzed using the IPA tool to determine where performance and/or resourcing of PGI attributes were not congruent with the expectations of PGI users (generally in the form of over-servicing or under-servicing). The IPA demonstrated that a majority of PGI users perceived the study site to be high performing and were satisfied with many of the assessed attributes. The survey identified the potential for some improvement of the amenity and/or infrastructure installations at the site, as well as directing attention towards a more effective utilization of scarce resources. Optimizing the management of PGI spaces will enhance opportunities for individuals to obtain the physiological, psychological, and emotional benefits that arise from experiencing quality urban PGI spaces. This case study promotes the important contribution that high-quality PGI spaces, which include remnant and restored UN spaces, make to the development of resilient and sustainable urban centers.

Keywords: biophilic design; green infrastructure; Importance-Performance Analysis (IPA); public open space; re-naturing cities; urban nature; visitor satisfaction survey; resource rationalization

---

1. Introduction

Public green infrastructure (PGI) is becoming an increasingly important and necessary part of urban life. Public green infrastructure is recognized as a mediator towards the emerging global challenges of climate change, which include global warming, extreme weather events, enhanced urban heat island effect, as well as general trends of declining public health and wellbeing [1,2]. Due to the multiple management and research disciplines that PGI intersects, a number of definitions and interpretations appear in the literature [3–6]. Public green infrastructure can be referred to as networks of public open space, biofiltration installations, public trees, green walls, green roofs and the like [1,3]. The delineation of green infrastructure, as opposed to other categories of infrastructure (i.e., blue or grey), is accepted as being the way in which the infrastructure harmoniously, and simultaneously, delivers relief to social and environmental pressures [3]. Within this case study, PGI is restricted to describing green urban public open spaces (e.g., parks with or without urban nature (UN)) and the contribution those spaces can make to urban living.

With continued population growth in urban centers, such as Perth, Western Australia, the opportunities for creating new PGI is limited [7]. Urban lifestyles are becoming increasingly psychologically demanding, with adverse trends that are in part the result of individuals rapidly disengaging with nature and the surrounding environment [2,8]. Additionally, trends are being documented that show increased social disconnection between people and places, which can be considered to be cultural disconnection in this context [9,10], as well as disconnection between people and nature [8,11]. These negative trends have resulted in the protection of PGI sites being considered more critical than in the past [2,8]. The planning and management of these sites are largely at the behest of local and state government land owners and management authorities, at times in conjunction with each other, and occasionally with community members [12–14]. Hereafter, all those combinations of ownership and management arrangements are collectively referred to as land managers or just managers as appropriate.

In the pursuit of land managers successfully catering for the current and future psychological and physiological health needs, and general wellbeing of communities utilizing PGI spaces, a substantial effort is required to thoroughly understand the needs of PGI users, how existing PGI spaces are being used, PGI users perception of site performance, and how attributes of a PGI contribute to a successful and engaging PGI experience. After this information has been collected, analyzed, and consolidated in policy and management plans, the quality management of PGI spaces is likely to be enhanced in both the short and long term.

Life in a highly urbanized setting is often met with persistent noise, light, and tension. It can be devoid of nature, well-functioning ecosystems, and lack opportunities for individuals to experience nature [11,15]. As such, in highly urbanized settings, PGI is of increased importance as it allows individuals to recreate, socialize, exercise, engage with nature, and engage in necessary spiritual reflection, which are integral to the human psyche for general health and wellbeing [15]. It has consistently been shown, and is now widely accepted, that PGI spaces provide users with various psychological, physiological, and general wellbeing benefits at both an individual and community level [2,16]. Public green infrastructure provides city dwellers with the opportunity to engage with the natural environment and ecosystems, in line with the Biophilic Hypothesis popularized by Wilson [16]. The Biophilic Hypothesis asserts that individuals possess a deep biological need and desire to engage with nature in order to thrive as a species [16]. It is reported that without engagement with nature, the general health and wellbeing of communities begins to decline [1]. Therefore, engagement with nature and the natural environment is paramount for building and developing resilient urban communities [17]. Building resilience is an important goal and mounting requirement for cities facing emerging and morphing social, economic and environmental changes and/or challenges.

Designing PGI that meets the needs of local and wider urban communities requires careful and measured consideration. Resource saturation (offerings in excess of requirements) and resource rationalization (distribution of resources based on requirement) need to be considered to ensure opportunities are provided in an equally distributed way (equity planning), ultimately supporting and encouraging broad engagement of a range of demographic subgroups [18,19]. Needs can significantly differ among such demographic subgroups. Understanding the needs of local communities marks the first step in planning to effectively and equitably meet these needs [18]. In its many forms, PGI has the ability to cater to a wide range of PGI users pursuing a variety of activities; active, passive, social, and/or recreational. Ultimately, PGI user groups are responsible for determining the performance and efficacy in the ability of these spaces to meet their current and diverse needs. From the perceptive of a land manager, such information is valuable when seeking to deliver the psychological, physiological, and general well-being outcomes afforded by PGI spaces. Demographic and satisfaction information must be gathered through quality engagement and consultation among PGI users, stakeholder communities, and land managers.

A number of techniques may be utilized when interrogating the collected information as described above, pursuant to improving on-ground management of urban PGI. While techniques of Importance-Performance Analysis (IPA) are widely applied to determine how well the attributes of a good or service perform in meeting the expectations of consumers, clients, users, and visitors (e.g., [18,20–27], IPA has only rarely been applied to enhancing PGI management [18,28]. The literature review of Parker and Simpson [28,29] and additional research in support of this case study, identified only four other studies that have utilized IPA techniques to investigate how PGI users perceive urban PGI spaces [30–33]. The IPA is a tool that analyses two dimensions of an experience by comparing the importance of the attributes of a product or service, urban PGI in this context, with the performance of those attributes in meeting user expectations [23,27]. In this case study, IPA was utilized to assess the importance of attributes of high quality PGI identified from the literature [28,29] against their relative performance for an urban PGI space in Perth, Western Australia. This case study demonstrated how the IPA tool was able to express the relationship between these two dimensions of experience in order to determine user satisfaction with each PGI attribute. Informed by the IPA, it was possible to identify where performance and/or resourcing of PGI attributes were not congruent with the expectations of PGI users, generally in the form of over-servicing or under-servicing. As a relatively straightforward to use and easy to interpret primary indicator, IPA is an efficient way for land managers to survey PGI users to determine evidence-based management action.

2. Materials and Methods

2.1. Study Site

Perth is a picturesque city on the south-western coast of Australia. Perth has a current population of approximately 2.6 million people [34]. Framed by natural assets, such as the Swan and Canning rivers, hundreds of kilometers of coastline, remnant native vegetation, and ancient geological features, with a mild climate, and a politically and socially stable society, Perth is among the highest ranking cities for liveability [35,36]. The population and footprint of Perth continues to grow through a combination of densification directed towards infill development and sprawling development on the urban fringe [37,38].

The locale of Perth experiences a Mediterranean climate with the indigenous vegetation being a mix of *Banksia* and eucalyptus woodlands with some *Agonis* and Tuart Forest [8,39–42]. Prior to British colonization, 'Perth' and the surrounding landscape was an extensive network of wetlands that supported the first nation Noongar population for at least 45,000 years [8,43,44]. Since the arrival of the British in 1829, urban development has seen many of these wetlands filled in and/or destroyed [8,43]. Within a 20-kilometer radius of the Perth city center, several significant wetlands still exist however, including Lake Claremont, Lake Monger, Bold Park, and Herdsman Lake (Figure 1) and those wetlands also contribute to the identity of Perth as an urban center [8].

The chosen study site, Lake Claremont (31.9738° S, 115.7771° E), which is a PGI and UN space covering over 60 hectares, is located approximately 10 kilometers south-west of the Perth city center. Under environmental protection policies of the Government of Western Australia, Lake Claremont is included on the list of Swan Coastal Plain Conservation Category Wetlands, the lake and the adjacent PGI are zoned as a Bush Forever site, and an Environmentally Sensitive Area in terms of the remnant and restored native vegetation [8]. The Lake Claremont site has diverse offerings and includes formal and informal active and passive recreation opportunities, as well as remnant and restored UN spaces. The site includes a Par 3 golf course, playgrounds, open turf spaces, a dog exercise area, barbecue (BBQ) facilities, and major renaturing efforts in the form of revegetation with local native species [8].

2.2. Field Survey

Despite large-scale visitor satisfaction surveys pertaining to PGI sites being limited in the published literature, the recognized benefits of surveying are extensive [18,45,46]. Quality surveying of PGI spaces has been shown to improve decision-making capabilities; strengthen support for infrastructure installations, upgrades, removals and prioritization; facilitate better environmental, conservation, and educational outcomes; better meet the needs and desires of the PGI users; and enhance opportunities for mental and physical wellbeing [18,29,46].

Figure 1. Location of the Lake Claremont public green infrastructure (PGI) space in relation to adjacent green infrastructure. The Perth Central Business District is located immediately east–north-east of the Kings Park PGI space. Adapted from the Western Suburbs Regional Organisation of Councils (WESROC) Greening Plan 2002 [47].

The associated data paper by Simpson and Parker [46] provides a detailed description of the design and implementation of the survey reported by this article. The following paragraph summarizes the information provided in the data descriptor paper.

A convenience intercept survey was implemented at the Lake Claremont PGI space on several days in December 2016 and January 2017 to coincide the with peak holiday and recreation period of the Austral summer [18,22,48]. The anonymous pen and paper self-report questionnaire used for the survey had categorical demographic questions as well as the question and paired Likert scales required for an IPA assessing 22 attributes of quality PGI spaces identified from the literature [18,28,29,46]. The IPA asked PGI users *"How important are the following features of Lake Claremont to you and how satisfied are you with their management?"* Participants could provide their importance ranking for each of the 22 attributes related to the quality PGI space using a 5-point Likert scale that ranged from 1 = *Not At All Important* to 5 = *Extremely Important*. Participants provided their performance rankings using a modified 6-point Likert scale that started at 0 = *Unable To Report* and then spanned from 1 = *Not At All Satisfied* to 5 = *Extremely Satisfied*. See Simpson and Parker [46] for further discussion on the setting of these values for the Importance and Performance Likert scales.

2.3. Data Analysis

De-identified data from the n = 393 returned questionnaires were transposed to a Microsoft Excel spreadsheet. That data is available as a .csv file attached to the data descriptor of Simpson and Parker [45]. Microsoft Excel 2016 was also used to analyze and graph the data presented in this article.

While demographic data from all PGI users who participated in the survey is reported in the Results section below, only data from PGI users who could or choose to report on the performance of the attributes associated with the study site were included in the IPA. Hence, importance data from PGI users who provided no response or a zero score regarding the performance for an attribute was excluded from the IPA.

As per Oh [21] and Taplin [27], the assumption of a correlation between the importance and performance rankings were checked for each of the 22 PGI attributes assessed by determining the significance of the Pearson correlation coefficient [49]. The absence of a correlation between the importance and performance rankings for a PGI attribute suggests the potential for a difference in how subgroups of PGI users perceive the performance of that attribute [50,51].

Scale-Centered IPA (SC-IPA), Data-Centered IPA (DC-IPA), and Gap Analysis IPA (GA-IPA) were performed [20,27,52]. The SC-IPA is presented graphically as a grid with four quadrats (see Results, Figure 2) that are centered on the midpoints of the importance and performance scales, where the measures switch from not import or not satisfactory to being important or performing satisfactorily [22,27,53]. For this reason, it is critical that the Likert scales used to gather the importance and performance data are equivalent in the span of the measurement categories **and** that the midpoint of those scales is a neutral value [27,53] While the midpoint may be implicit for a Likert Scale with an even number of categories, a majority of researchers recommend using Likert scales with an odd number categories with the neutral value being explicit in the response options provided to participants [18,22,53–56]. In scenarios where IPA attributes are assessed as high performing on the SC-IPA, then the enhanced DC-IPA and GA-IPA can provide insights that may facilitate management actions to address attributes that may have declining performance or attributes that are being over-serviced and are therefore consuming scarce resources with no perceived benefit for PGI users [18,22,27].

Similar to the SC-IPA, the DC-IPA is also presented graphically as a grid with the four quadrats *Keep Up Good Work, Focus For Management, Low Priority, Possible Over-servicing*. The grid for the DC-IPA is however centered on the grand mean of the importance and grand mean of the performance of all the attributes assessed [22,53,57]. Focusing the grid on the means of the importance and performance ratings provides greater clarity for the prioritization of management action on those attributes that may be performing below the expectations of PGI users, especially for high-quality PGI that may have all attributes located in the Keep Up Good Work quadrat of the SC-IPA.

The GA-IPA was completed by determining the differences between the performance and importance rankings (i.e., Gap = Performance − Importance) by each participant for each attribute, checking that the performance gaps for all participants approximated a normal distribution for each attribute, calculating the mean of the gaps for each PGI attribute, and checking if the mean gap for all responses was significantly different from zero (i.e., Performance $\neq$ Importance) using a one sample t-Test [22,23,27]. Consistent with to the approach of Smolčić Jurdana et al. [24–26] and Taplin [27], this study presents both the IPA grid for the DC-IPA and the GA-IPA on the same figure provided in the Results. The greater the gap in the performance of an attribute, which is represented by its distance from the line of parity (Performance = Importance) in the graphical representation, the higher the management priority to implement corrective action [18,22]. The graphical prioritization of management actions can be further enhanced by combining the grids from the DC-IPA and GA-IPA, as demonstrated in Parker [18] and Taplin [27].

Soldić Frleta [23] and Taplin [27] report that the larger the gap is between the importance and performance rankings, the lower the satisfaction is likely to be with that PGI attribute. With additional rationalization, management action can then focus on improving the performance of attributes with a negative gap in order to meet the expectations of PGI users. Attributes with positive gaps, where performance exceeds the expectation of PGI users, could also require action to reduce over-servicing and more effectively allocate scarce resources available for PGI management.

3. Results

3.1. Demographic Data

The demographic profile of the study site users in the Austral summer of 2016–2017 is shown in Table 1. The approximately 2:1 ratio between PGI users who identified as female or male is consistent with ocular-based gender counts performed during the survey.

Table 1. Demographic profile for PGI users at Lake Claremont (n = 393) in the Austral summer of 2016–2017.

Gender Profile		
Categories	Responses	Percentage $\pm$ 95% Confidence Interval
Female	241	61.3 $\pm$ 4.8
Male	144	36.6 $\pm$ 4.8
Other	3	0.8 $\pm$ 0.9
Prefer not to Disclose	1	0.2 $\pm$ 0.5
No Response	4	1.0 $\pm$ 1.0
Age Profile		
Categories	Responses	Percentage $\pm$ 95% Confidence Interval
18-24	17	4.3 $\pm$ 2.0
25-34	29	7.8 $\pm$ 2.6
35-44	64	16.3 $\pm$ 3.6
45-54	95	24.2 $\pm$ 4.2
55-64	85	21.6 $\pm$ 4.1
65+	99	25.2 $\pm$ 4.2
No Response	4	1.0 $\pm$ 1.0
Usual Place of Residence		
Categories	Responses	Percentage $\pm$ 95% Confidence Interval
Surrounding Suburbs (< 5 km)	305	77.6 $\pm$ 4.1
Other Metropolitan Suburbs	66	16.8 $\pm$ 3.7
Regional Western Australia	8	2.0 $\pm$ 1.4
Other Australian States	9	2.3 $\pm$ 1.5
International	5	1.3 $\pm$ 1.1

3.2. Importance Performance Analysis

The quantitative values for the IPA are provided in Table 2 and in Figures 2 and 3. With the exception of Attribute 8* (*High-quality European-/English-themed spaces and areas*), users of the study site perceived attributes to be performing in the Keep Up Good Work quadrant (Figure 2) of the SC-IPA recommended by Martilla and James [20]. Having the analyzable attributes (excluding Attribute 8*) located in the top right quadrant of the SC-IPA demonstrates that the study site is, overall, perceived to be a high quality PGI space that is meeting the expectations of the majority of PGI users.

With 120 unanalyzable responses (30.5%), Attribute 8* had the highest rate of invalid or blank responses for the 22 attributes tested by the IPA questions and, while significant (p = 0.0409), there was only weak correlation between the importance and performance rankings for that attribute (12%). In addition, that attribute created a high degree of confusion and questioning by survey participants. For those reasons, Attribute 8* was excluded from the enhanced IPA analyses reported for this study.

Table 2. The importance of attributes of quality PGI spaces for users of the Lake Claremont site and the performance of those attributes with respect to meeting user expectations (i.e., visitor satisfaction). Attribute numbers relate to the data labels used in the Importance-Performance Analysis (IPA) plots (Figures 2 and 3).

Attribute Number	Attribute	n	Mean Imp.	Mean Per.	Sig. Corr.	Gap (P-I)	Sig. Gap
1	Availability of shade—Trees or Structures	357	4.20	3.89	0.1379	−0.31	***
2	Bird watching infrastructure—Observation Deck, Rotunda	310	3.38	3.68	***	0.31	***
3	Children's playground(s)	306	3.55	3.69	***	0.14	0.0612
4	Directional signs within the park	312	3.24	3.54	0.2978	0.30	***
5	Dog exercise area	306	3.79	3.24	0.0556	−0.55	***
6	Ease of access to and around site	339	4.12	4.03	***	−0.10	0.0702
7	Fencing	323	3.36	3.62	***	0.26	***
8*	High-quality European/English themed spaces and areas	273	2.63	3.50	*	0.87	***
9	High-quality infrastructure—Paths, Lights, Toilets, barbecues (BBQs), Benches	347	3.82	3.33	0.0834	−0.49	***
10	High-quality lake water body	341	4.30	3.55	***	−0.75	***
11	High-quality nature spaces and areas	338	4.42	3.88	***	−0.54	***
12	High-quality services—Café, Gym, Golf Club	321	3.66	3.69	***	0.03	0.6386
13	High-quality turf	316	3.29	3.65	***	0.36	***
14	Interpretive information and signs	326	3.48	3.48	**	0.00	0.9599
15	Native fauna presence and activity	343	4.41	3.87	***	−0.54	***
16	Off-leash dog exercise	300	3.71	3.01	***	−0.70	***
17	On-leash dog walking	307	3.77	3.71	***	−0.07	0.4174
18	Other sporting installations—Aquatic Center, Cricket, Hockey, Tennis	305	3.50	3.75	***	0.25	***
19	Par 3 Golf Course	281	2.98	3.44	***	0.46	***
20	Park exercise equipment	316	3.42	3.58	***	0.15	*
21	Personal safety	347	4.29	3.90	0.5912	−0.38	***
22	Tree management	341	4.39	3.89	***	−0.50	***

n = Sample Size = Number of analyzable responses. Mean Imp. = Mean value of the importance rankings for that attribute. Mean Per. = Mean value of the rankings of visitor satisfaction with the performance of that attribute. Sig. Corr. = Outcome for the test for statistical significance of the Pearson correlation between importance and performance rankings. Gap = Mean Performance − Mean Importance. Sig. Gap = Outcome for statistical test for significance of the Gap being > 0 (i.e., Performance ≠ Importance). Outcomes of statistical test are reported as the *p*-value or as * for *p* < 0.05, ** for *p* < 0.01, or *** for *p* < 0.001.

Even the preliminary SC-IPA level of analysis provides additional insights for the management of the study site. Positioned on the quadrant boundaries, performance of Attribute 16 (Off-leash dog exercise) and Attribute 19 (Par-3 Golf Course) could easily slide into the Focus For Management quadrant and the Possible Over-servicing quadrant respectively [22,27]. Further insights relating to those two attributes are provided by the DC-IPA and GA-IPA presented below.

As noted in the Methods section, the enhanced DC-IPA and GA-IPA can assist managers of high-quality PGI sites to prioritize management actions to address attributes that are performing below the expectations of PGI users and to review the allocation of scarce resources to attributes that PGI users perceive as being over-serviced. The combination of the DC-IPA and GA-IPA approaches presented in Figure 3 highlights that a number of the attributes of the study site are meeting the expectations of PGI users with an appropriate level of resourcing. The enhanced IPA provided in Figure 3 also provides other insights for the study site managers that are explored below.

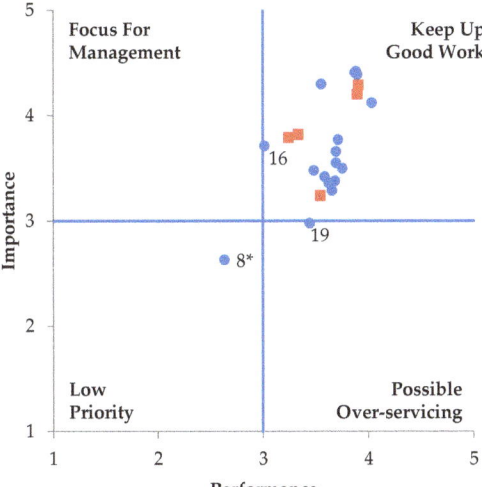

Figure 2. Scale-centered IPA. Attributes shown as a red square if there was not a significant correlation between the Importance and Performance ratings, which suggests possible differences in opinion between subgroups of PGI users regarding that attribute.

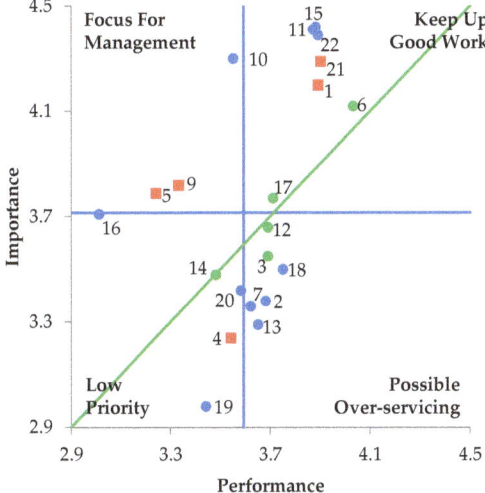

Figure 3. Combination Data-Centered and Gap Analysis Importance-Performance Analysis. Attributes are shown as a red square shown if there was not a significant correlation between the Importance and Performance ratings. For attributes shown as a green dot Performance = Importance. Attributes shown as a blue dot have non-zero gaps and Performance ≠ Importance.

The GA-IPA demonstrates that the study site users are satisfied with the performance and resourcing of Attribute 14 (*Interpretive information and signs*), Attribute 3 (*Children's playground(s)*), Attribute 12 (*High-quality services—Café, Gym, Golf Club*), Attribute 17 (*On-leash dog walking*), and Attribute 6 (*Ease of access to and around the site*). Although, with a p-value of 0.0612 and a location in the lower right Possible Over-servicing quadrant of the DC-IPA, there is potential for the provision and servicing of children's playgrounds at the Lake Claremont site to exceed the need perceived by the community.

The upper left Focus For Management quadrant of the DC-IPA reveals that a majority of users of the study site perceive that Attribute 10 (*High-quality lake water body*) as performing below average and having the largest gap of the GA-IPA (−0.75) is evidence that the majority of PGI users perceive the water quality of Lake Claremont to be the worst performing site attribute and the attribute most in need of management action.

Remaining focused on the upper left quadrant of the DC-IPA also reveals that Attribute 5 (*Dog exercise area*) is partnered with Attribute 16, meaning that two aspects of dog management at the study site are performing significantly below the expectations of PGI users. Further, off-leash dog walking (Attribute 16) with a gap of −0.70 is the second worst performing attribute of the study site, after the perceived quality of the lake waterbody. In addition, there was not a significant correlation between the importance and performance ratings for Attribute 5, which suggests possible differences in opinion between subgroups of PGI users regarding management of dog exercise at the study site. The responses for Attributes 5 and Attribute 16 are in stark contrast with the uniform satisfaction of all PGI users with on-lead dog walking at the study site (Attribute 17).

Attribute 9 (*High-quality infrastructure—Paths, Lights, Toilets, BBQ, Benches*) is also position in the top left Focus For Management quadrant of the DC-IPA, is located significantly above the line of parity for the GA-IPA, and also lacks correlation between the importance and performance rankings of PGI users, hence that attribute should also be a focus for management to determine what elements related to the broad scope of PGI infrastructure are perceived to be underperforming by the subgroups who use the study site.

For GA-IPA related to commercial operations, the management focus is always directed towards attributes located significantly above the line of parity, where the expectations of paying customers are not being met. The rationale for this focus being that the larger the gap between importance and performance of those attributes, then the lower the customer satisfaction with their experience [23,27]. In contrast, the management of PGI spaces requires the balancing of scare resources against the service expectations of PGI users. For that reason, we recommend that PGI managers next consider those attributes below the line of parity, particularly those attributes that are located in the bottom right Possible Over-servicing quadrant of the DC-IPA, which are generally ignored in most IPA studies. That review may identify scare resources that can be reallocated to address poorly performing attributes located significantly above the line of parity.

As previously mentioned, the level of service provided by the Par 3 Golf Course (Attribute 19) was the attribute of the study site that PGI user perceived to be most over-serviced with the largest positive gap (0.46) between perceived performance and its importance rankings. In priority order for management consideration, based on PGI user perceptions of over-servicing, the following attributes should be reviewed: Attribute 13 (*High-quality turf*), Attribute 2 (*Bird watching infrastructure—Observation Deck, Rotunda*), Attribute 4 (*Directional signs within the park*), Attribute 7 (*Fencing*) and Attribute 18 (*Other sporting installations—Aquatic Center, Cricket, Hockey, Tennis*). Reviewing the potential over-servicing regarding the directional signs in the park (Attribute 4) would be complicated by the lack of correlation between the importance and performance rankings, suggesting different perceptions of the signage among subgroups of PGI users.

Having investigated the potential to reallocate resources from attributes that may be over-serviced, the land managers should return their attention to the attributes located in the upper right Keep Up Good Work quadrant that are significantly above the line of parity. After the water quality of the study site (Attribute 10), then the worst performing attributes in priority order under the GA-IPA are Attribute 15 (*Native fauna presence and activity*), Attribute 11 (*High-quality nature spaces and areas*), and Attribute 22 (*Tree management*). The significant correlation between the importance and performance ratings for those three attributes provides evidence that a majority of PGI users at the study site perceive the management of those attributes to be underperforming. Attribute 21 (*Personal safety*) and Attribute 1 (*Availability of shade—Trees or Structures*) are both preforming significantly below PGI user expectations, but again the lack of correlation between the importance and performance rankings for these attributes, is evidence for a difference of opinion between subgroups of PGI users at the study site, which would complicate the management actions required to address the perceived underperformance in those attributes.

4. Discussion

4.1. Demographic Profile of Public Green Infrastructure (PGI) Users

Researchers commonly report that understanding PGI user experiences, expectations, and satisfaction levels is of great value to inform the actions and decisions of land managers [58–61]. Meeting physical, psychological, spiritual, or other community needs and also providing abundant social, economic, and environmental opportunities are primary services delivered by urban PGI spaces [59]. Creating and enhancing the synergy between PGI users and land managers is critical to improved site management. While adjusting and adapting to the evolving needs and desires of PGI users is difficult, doing so is, however, confirmed as a best practice approach.

The age distribution of the survey participants for the PGI space investigated in this study did not match the current local or regional distributions reported by the Australian Bureau of Statistics and summarized in Tables S1 to S11 that are provided as supplementary material [62,63]. The age of the survey participants was moderately skewed towards an older population. The implications of the age distribution should be considered in the planning and management of PGI spaces. Researchers such as Johnson and Glover [61] suggest that passive-park attributes such as resting spaces, viewing infrastructure, and attributes that support flora and fauna experiences are more likely to draw visits from older PGI users. A known prevalence of older PGI users also requires consideration of attributes such as safe access, correctly graded paths, support rails, and more passive park attributes that may require higher levels of service or prioritization. There may also be less demand for attributes like playgrounds that engage younger PGI users and, therefore, such attributes could require a lower level of service.

Similarly, the gender distribution of the surveyed population was moderately skewed towards females. The gender distribution also did not match current local or regional distributions [62,63], which potentially indicates different values and choices of PGI users towards leisure/recreation activities and other attributes of the study site. A similar trend has been reported in studies such as Siu et al. [64]. As for the skewed age distribution, knowing a higher number of females visit the site has implications for the site managers. Items such as sense of safety, lighting, seating, and other infrastructure may require higher levels of service [64]. Understanding the demographic information of urban PGI users is valuable on several levels and this value can be realized through the hierarchy of benefits listed in Table 3.

Table 3. Hierarchy of benefits to be gained from understanding the demographics of PGI users.

Benefits to be Gained
Ensuring the strategic direction and future planning of the PGI by the land manager is aligned and congruent with the site users.
Considering the current strategic direction for the PGI and to better allow for estimates of future PGI user demographics.
Assessing proposals for infrastructure installations, upgrades, removals and prioritization.
Assessing maintenance/operating budgets and their ability to service the needs and desires of the site users.
Creating the basis for further investigations, such as quantifying and qualifying the importance and satisfaction levels of the site users.

4.2. Outcomes of Importance-Performance Analysis (IPA)

While prioritization and implementation of management actions related to any PGI space requires a holistic approach that incorporates other factors, such as financial, social, cultural and political implications and constraints as well as ecological values and environmental services, in the decision-making processes, the key findings of the IPA reported in this case study can be summarized as follows.

While potentially a difficult issue to address, because of differing views among subgroups of PGI users, dog exercise and off-leash dog walking at the study site should be a priority for management action as both those attributes were performing significantly below the expectations of PGI users. Those findings are in stark contrast to on-leash dog walking at the site, which a majority of PGI users perceive to be satisfactory.

It may be possible to reallocate funding and other scarce management resources from attributes that PGI users perceive to be over-serviced to those attributes that PGI users report to be underperforming. Attributes at the study site that PGI users perceived to be over-serviced (i.e., have significant positive gaps) are the Par-3 golf course and other sporting installations, the bird watching rotunda and observation deck, and the conservation and/or safety fencing at the site. Potentially, the playgrounds for children could also enter this category and any future allocation of resources to that attribute would need to be carefully considered.

The four most poorly performing attributes in terms of the GA-IPA (i.e., largest negative gaps) are related to the environmental values and services of the site. The quality of the lake waterbody, the presence and activity of native fauna, the quality of the nature spaces, and tree management should all receive additional management action to ensure that the performance of those attributes meets the expectations of PGI users. After the combination of dog exercise and off-leash dog walking, addressing the real or perceived poor quality of the lake waterbody should be the next highest priority for management action. While significantly underperforming and in need of action, the other three environmental attributes are all located in the Keep Up Good Work quadrant of the DC-IPA, which is fortuitous as improving the performance of those three attributes is likely to require consistent, focused effort for an extended period of time, spanning into years, if not decades [65,66].

4.3. Changing Values and Expectations

While Australia is a wealthy nation, it has only been in recent times that the community has come to place a relatively high value on natural assets [8]; however, the expectations of PGI users reported in the literature [8,29,58] are generally being met by the study site. The expectations of PGI users towards the attributes such as interpretive information and signs, children's playgrounds, high-quality services like café and gym, and on-leash dog walking are being fully met. Based on discussions and interactions with and responses from survey participants, some emerging and changing values were observed at the study site. These included the desire for increased protection of local native vegetation, enhancement of lake water quality, the desire to increase opportunities to view the lake, enhancing the quality and protection of indigenous fauna, removal of ecological threats, protection and retention heritage markers, as well as a balanced approach to dog management. These findings are consistent with the changing values reported by Jones and Newsome [35] and Simpson and Newsome [8]. Some of these emerging community values present complex challenges based on the individual positions and the potential for conflicting expectations and perceptions of PGI users (e.g., protection of native vegetation and enhancement of views to the lake body requiring removal of renatured vegetation). Values may also be difficult to quantify and measure, which can result in difficulty assessing their ongoing performance. To understand emerging and changing values across PGI spaces, targeted site-specific research is required to further quantify the preliminary findings of this case study.

4.4. Land Manager Responsibilities

This case study demonstrates the need for land managers to adopt a flexible and evolving approach when managing urban PGI spaces. This will facilitate better harnessing of opportunities, supporting quality engagement of PGI users, and result in a reduction of resources required to yield the same (perceived) quality space. This approach to management will also allow PGI spaces to improve in performance and offerings for users. Land managers should be cautious not to become complacent with well performing spaces, rather opportunities should always be sought to improve urban PGI spaces, commensurate with the valuable community assets that they are. These opportunities are most likely to arise from engagement with the community.

4.5. Frequency of Surveying

With the intensification of PGI user needs, desires, and expectations, land managers must respond quickly, with an evidence-based approach, if the quality and value of PGI spaces is to be retained and improved. Frequency of surveying PGI users must be determined to maintain a confident understanding of the expectations of PGI users and satisfaction levels with a PGI space. The published literature is largely silent on frequency for surveying PGI satisfaction; however, informed by the literature review of Parker and Simpson [28,29], some factors to consider when determining frequency are provided in Table 4. After considering these factors (as well as other site specific factors), a suitable survey frequency can be determined. A suitable time period between surveys is generally considered to be between two and five years. The changing status of the PGI space itself as well as the urban landscape in which the PGI space exists may also be responsible for increasing the frequency of surveying.

Table 4. Factors to be considered in determining a suitable survey frequency to facilitate evidence based management of urban PGI spaces.

Determinants of Survey Frequency
1. Development within and around the site, including new installations, upgrades, removal of assets.
2. Maturity of the space; spaces in their infancy would likely require an increased surveying frequency as opposed to more mature spaces.
3. Changes in the patronage of the site—a substantial increase or decrease in patronage (i.e., 20%) can be considered cause for increased surveying frequency.
4. Changes in political pressures or support for the space, particularly those that could affect (financial or otherwise).
5. Resource changes, including grant opportunities or changes in financial constraints.
6. Social pressures, including changes in social values.
7. Future planning around development or re-zoning.
8. Availability of skills, funding and other resource required for future surveying.

Source: Informed by systematic literature review of Parker and Simpson [20,21].

4.6. Limitations and Lessons

This case study demonstrates that from the perspective of a land manager, IPA can reveal much information and provide direction when attempting to correct resourcing misallocations, generally in the form of over-servicing and/or underperformance/under-servicing of attributes that are failing to meet expectations of PGI users. While the majority of the current IPA literature (e.g., [21,23,27]) recommends that management action be directed to underperforming attributes before considering over-serviced items, this may not give the complete picture in the case of PGI management, because resourcing is generally scarce. Considerations should also be made towards the original financial investment and maintenance requirements of each attribute of the PGI space when prioritizing management actions. For example, an attribute may deviate from the line of parity substantially, but may not attract a large financial investment, and therefore a case can be made to balance the deviation from the line of parity in conjunction with the level of financial investment to determine correct order for management action. Such was the rationale for a lower prioritization for management action in relation the attributes associated with the nature spaces, native fauna, and trees provided previously in Section 4.2 of this Discussion. This balanced approach will produce a more reliable and robust methodology when prioritizing management action (Figure 4). Such an approach is also likely to gain financial savings more rapidly, which can be reallocated to expedite the correction of underperforming attributes more readily.

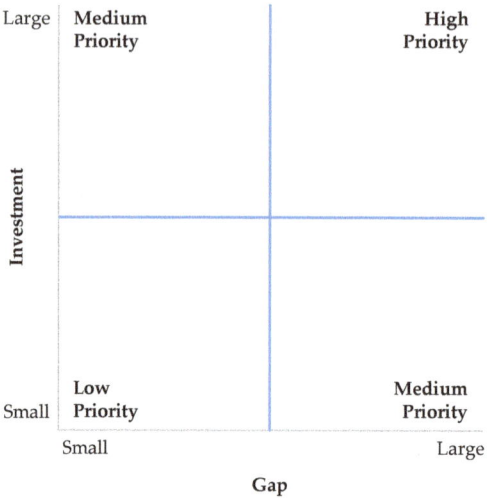

Figure 4. Conceptual model for the prioritization of corrective management action by balancing investment in an attribute and the gap in performance with respect to the expectations of PGI users.

4.7. Further Research

The case study and IPA findings reported by this article assume that all PGI attributes impact equally upon the experience of a PGI user. In reality, different attributes may affect different PGI users to different degrees. This could be further developed with research to determine the relative weighting of attributes for subgroups of PGI users in an attempt to account for differences in impact for common PGI attributes.

Further research utilizing IPA on PGI assets could include temporal analysis of repeat IPA surveys in an attempt to identify trends for issues emerging at the site, allowing for land managers to address the issues in their infancy before they become more complex and expensive to address.

It is important to consider what PGI attributes PGI users choose to engage with while on site. Such information is important when considering the availability, service levels, and opportunities provided by the attributes of high-use areas within PGI spaces. Further research with respect to how people utilize the study site would provide insight into any differences in the values, satisfaction, and perceptions of attribute and management performance among different subgroups of PGI users. Information on activity engagement is also useful when undertaking cost-benefit analysis of future infrastructure proposals or installations.

In addition, it has been observed within this case study and the supporting literature review of Simpson and Parker [28,29] that the PGI user needs have intensified, thus alternate or more detailed performance analysis may be required in order to better understand the implications there in. A planned program of repeat surveys at intervals of three to five years would contribute to that understanding.

5. Conclusions

Public green infrastructure in urban centers is invaluable. These spaces are invaluable because they provide the opportunity for PGI users to connect with and experience nature, as a locally accessible asset and experience that cannot be found in any other way. Experiencing and connecting with nature provides humankind with positive psychological and physiological benefits and spiritual well-being. Which are facets of citizen life that are being lost with each generation as humanity and the planet become ever more urbanized.

Given the benefits and soulful contribution that resourcing PGI provides for the betterment of the local community and society as a whole, it is even more important to appropriately allocate, protect, rationalize, and maximize the return of these resources. The IPA tool has not been widely utilized in PGI planning and management, however this case study demonstrates its applicability to inform management and maximize the return on scarce PGI resource allocations. This case study further demonstrates how IPA can assist in the prioritization and rationalization of resources and the drive that land managers can show towards equity planning for the wider community to access PGI assets.

For each of the 22 attributes of the Lake Claremont PGI space, the majority of which are directly transferrable to other urban PGI spaces, this case study revealed those attributes that were meeting the expectations of PGI users, those attributes that PGI users perceived to be underperforming, and those attributed that were or had the potential to be over-serviced. In contrast to other applications of IPA, this PGI-focused case study promotes land managers investigating attributes that PGI users perceive as being over-serviced to determine the potential for generally scarce resources to be reallocated to improve underperforming attributes. Furthermore, the correlation analysis utilized in this case study revealed several attributes of the Lake Claremont PGI space about which subgroups of PGI users may have different perceptions regarding the level of performance of those attributes and recommends that land managers undertake additional investigation before deciding on any course of action regarding those attributes.

This research advocates equity planning in all demographic ranges having access to PGI spaces that meets local needs. This quality, regardless of the PGI attributes desired, should be somewhat comparable, regardless of social and economic status, population density, and management affluence. Importance-Performance Analysis is one tool that assesses and confirms the performance of PGI attributes and thus supports equity planning through suggestion of resource need and rationalization.

Supplementary Materials: The following are available online at http://www.mdpi.com/2073-445X/7/4/159/s1, Tables S1 to S11 Demographic Data.

Author Contributions: J.P. and G.S. made equal contributions to this paper and as such are co-first authors.

Funding: This research received no external funding.

Acknowledgments: We thank our colleague David Newsome for his guidance on the Masters research by J.P. and comments on the associated thesis. We thank Dianne Parker for her efforts in transcribing the anonymous survey data from the self-report questionnaires to an Excel spreadsheet. We also give thanks to the Guest Editor of the *Land Urbanism and Green Infrastructure* special issue, and two anonymous reviewers whose comments enhanced our article. We would like to give particular thanks to *Land* Editorial Team for their informative, professional, and timely assistance that facilitated the publication of our article. This research was undertaken under Murdoch University Human Ethics Committee Approval 2016/213.

Conflicts of Interest: The authors declare no conflict of interest.

References

1. Beatley, T. *Biophilic Cities: Integrating Nature into Urban Design and Planning*; Island Press: Washington, DC, USA, 2011; ISBN 978-1-5972671-5-1.

2. Keniger, L.E.; Gaston, K.J.; Irvine, K.N.; Fuller, R.A. What are the benefits of interacting with nature? *Int. J. Environ. Res. Public Health* **2013**, *10*, 913–935. [CrossRef] [PubMed]

3. Norton, B.A.; Coutts, A.M.; Livesley, S.J.; Harris, R.J.; Hunter, A.M.; Williams, N.S. Planning for cooler cities: A framework to prioritise green infrastructure to mitigate high temperatures in urban landscapes. *Landsc. Urban Plan.* **2015**, *134*, 127–138. [CrossRef]

4. Chini, C.M.; Canning, J.F.; Schreiber, K.L.; Peschel, J.M.; Stillwell, A.S. The green experiment: Cities, green stormwater infrastructure, and sustainability. *Sustainability* **2017**, *9*, 105. [CrossRef]

5. Demuzere, M.; Orru, K.; Heidrich, O.; Olazabal, E.; Geneletti, D.; Orru, H.; Bhave, A.G.; Mittal, N.; Feliu, E.; Faehnle, M. Mitigating and adapting to climate change: Multi-functional and multi-scale assessment of green urban infrastructure. *J. Environ. Manag.* **2014**, *146*, 107–115. [CrossRef] [PubMed]

6. Sussams, L.W.; Sheate, W.R.; Eales, R.P. Green infrastructure as a climate change adaptation policy intervention: Muddying the waters or clearing a path to a more secure future? *J. Environ. Manag.* **2015**, *147*, 184–193. [CrossRef] [PubMed]
7. Grose, M.J. Changing relationships in public open space and private open space in suburbs in south-western Australia. *Landsc. Urban Plan.* **2009**, *92*, 53–63. [CrossRef]
8. Simpson, G.; Newsome, D. Environmental history of an urban wetland: From degraded colonial resource to nature conservation area. *GEO Geogr. Environ.* **2017**, *4*, 1–18. [CrossRef]
9. Soga, M.; Yamaura, Y.; Aikoh, T.; Shoji, Y.; Kubo, T.; Gaston, K.J. Reducing the extinction of experience: Association between urban form and recreational use of public greenspace. *Landsc. Urban Plan.* **2015**, *143*, 69–75. [CrossRef]
10. Patroni, J.; Day, A.; Lee, D.; Chan, J.K.L.; Kerr, D.; Newsome, D.; Simpson, G.D. Looking for evidence that place of residence influenced visitor attitudes to feeding wild dolphins. *Tour. Hosp. Manag.* **2018**, *24*, 87–105. [CrossRef]
11. Miller, J.R. Biodiversity conservation and the extinction of experience. *Trends Ecol. Evol.* **2005**, *20*, 430–434. [CrossRef]
12. Newsome, D.; Moore, S.A.; Dowling, R.K. *Natural Area Tourism: Ecology: Impacts and Managemen*; Channel View Publications: Bristol, UK, 2012; pp. 251–254, ISBN 978-1-84541-381-1.
13. Kellert, S.R.; Mehta, J.N.; Ebbin, S.A.; Lichtenfeld, L.L. Community natural resource management: Promise, rhetoric, and reality. *Soc. Nat. Resour.* **2000**, *13*, 705–715. [CrossRef]
14. Young, R.F.; McPherson, E.G. Governing metropolitan green infrastructure in the United States. *Landsc. Urban Plan.* **2013**, *109*, 67–75. [CrossRef]
15. Gladwell, V.F.; Brown, D.K.; Wood, C.; Sandercock, G.R.; Barton, J.L. The great outdoors: How a green exercise environment can benefit all. *Extreme Physiol. Med.* **2013**, *2*, 3. [CrossRef] [PubMed]
16. Wilson, E.O. *Biophilia*; Harvard University Press: Cambridge, MA, USA, 1984; ISBN 978-0-6740744-2-2.
17. Rupprecht, C. Informal urban green space: Residents' perception, use, and management preferences across four major Japanese shrinking cities. *Land* **2017**, *6*, 59. [CrossRef]
18. Parker, J. A Survey of Park User Perception in the Context of Green Space and City Liveability: Lake Claremont, Western Australia. Master's Thesis, Murdoch University, Perth, Western Australia, November 2017. Available online: http://researchrepository.murdoch.edu.au/id/eprint/40856/ (accessed on 15 December 2018).
19. Wolch, J.R.; Byrne, J.; Newell, J.P. Urban green space, public health, and environmental justice: The challenge of making cities 'just green enough'. *Landsc. Urban Plan.* **2014**, *125*, 234–244. [CrossRef]
20. Martilla, J.A.; James, J.C. Importance-Performance Analysis. *J. Mark.* **1977**, *41*, 77–79. [CrossRef]
21. Oh, H. Revisiting importance–performance analysis. *Tour. Manag.* **2001**, *22*, 617–627. [CrossRef]
22. Patroni, J. Visitor Satisfaction with a Beach-Based Wild Dolphin Tourism Experience and Attitudes to Feeding Wild Dolphins. Honours Thesis, Murdoch University, Perth, Western Australia, 2018. Available online: http://researchrepository.murdoch.edu.au/id/eprint/41944/ (accessed on 29 October 2018).
23. Soldić Frleta, D. Shifts in tourists' attitudes towards the destination offering. *Tour. Hosp. Manag.* **2018**, *24*, 020201. [CrossRef]
24. Smolčić Jurdana, D.; Soldić Frleta, D. Sustainable Rural Tourism Development-Tourists' Satisfaction with Istria as a Rural Holiday Destination. In Proceedings of the Biennial International Congress, Tourism & Hospitality Industry, Opatij, Croatia, 3–5 May 2012; University of Rijeka, Faculty of Tourism and Hospitality Management: Opatij, Croatia, 2012.
25. Smolčić Jurdana, D.; Soldić Frleta, D. Economic and Social Aspects of Rural Tourism. In Proceedings of the 6th International Conference, an Enterprise Odyssey: Corporate Governance and Public Policy–Path to Sustainable Future, Šibenik, Croatia, 13–16 June 2012; University of Zagreb, Faculty of Economics & Business: Zagreb, Croatia, 2012.
26. Smolčić Jurdana, D.; Soldić Frleta, D.; Župan, D. Assessment of Destination's Tourism Offering in the Off-Season. In Proceedings of the 4th International Scientific Conference-Tourism in Southern and Eastern Europe 2017: Tourism and Creative Industries: Trends and Challenges, Opatija, Croatia, 4–6 May 2017; University of Rijeka, Faculty of Tourism and Hospitality Management: Opatij, Croatia, 2017.
27. Taplin, R.H. Competitive importance-performance analysis of an Australian wildlife park. *Tour. Manag.* **2012**, *33*, 29–37. [CrossRef]

28. Simpson, G.; Parker, J. Data on Peer Reviewed Papers about Green Infrastructure, Urban Nature, and City Liveability. *Data* **2018**, *3*, 51. [CrossRef]

29. Parker, J.; Simpson, G. Public Green Infrastructure Contributes to City Liveability: A Systematic Quantitative Review. *Land* **2018**, *7*. in press.

30. Deng, J.; Pierskalla, C.D. Linking Importance–Performance Analysis, Satisfaction, and Loyalty: A Study of Savannah, GA. *Sustainability* **2018**, *10*, 704. [CrossRef]

31. Tonge, J.; Moore, S.A. Importance-satisfaction analysis for marine-park hinterlands: A Western Australian case study. *Tour. Manag.* **2007**, *28*, 768–776. [CrossRef]

32. Wang, Y.C.; Lin, J.C.; Liu, W.Y.; Lin, C.C.; Ko, S.H. Investigation of visitors' motivation, satisfaction and cognition on urban forest parks in Taiwan. *J. For. Res.* **2016**, *21*, 261–270. [CrossRef]

33. Yu, B.; Che, S.; Xie, C.; Tian, S. Understanding Shanghai Residents' Perception of Leisure Impact and Experience Satisfaction of Urban Community Parks: An Integrated and IPA Method. *Sustainability* **2018**, *10*, 1067. [CrossRef]

34. Australian Bureau of Statistics, 3218.0—Regional Population Growth, Australia, 2016–17. Available online: http://www.abs.gov.au/ausstats/abs@.nsf/mf/3218.0 (accessed on 31 October 2018).

35. Jones, C.; Newsome, D. Perth (Australia) as one of the world's most liveable cities: A perspective on society, sustainability and environment. *Int. J. Tour. Cities* **2015**, *1*, 18–35. [CrossRef]

36. The Economist Intelligence Unit. A Summary of the Liveability Ranking and Overview. Available online: http://pages.eiu.com/rs/783-XMC-194/images/Liveability_August2016.pdf (accessed on 16 October 2017).

37. Lehmann, S. The challenge of transforming a low-density city into a compact city: The case of the City of Perth, Australia. In *Growing Compact*; Bay, J.H.P., Lehman, S., Eds.; Routledge: Abindon, UK, 2017; pp. 95–119, ISBN 918-1-138-68040-1.

38. Khan, S.; Carville, A. To follow the Australian dream or to embrace urban densification. In *Growing Compact*; Bay, J.H.P., Lehman, S., Eds.; Routledge: Abindon, UK, 2017; pp. 301–316, ISBN 918-1-138-68040-1.

39. Beard, J.S. Definition and locations of the Banksia woodlands. *J. R. Soc. West. Aust.* **1989**, *71*, 85–86.

40. Simpson, G.D. Cracking the Niche: An Investigation into the Impact of Climatic Variables on Germination of the Rare Shrub *Verticordia staminosa* Subspecies *staminosa* (Myrtaceae). Honours Thesis, Murdoch University, Perth, Western Australia, 2011. Available online: http://researchrepository.murdoch.edu.au/id/eprint/8485/ (accessed on 25 October 2018).

41. Fowler, W. Soil Seed Bank Dynamics in Transferred Topsoil: Evaluating Restoration Potentials. Honours Thesis, Murdoch University, Perth, Western Australia, 2012. Available online: http://researchrepository.murdoch.edu.au/id/eprint/13389/ (accessed on 25 October 2018).

42. Ritchie, A.; Sinclair, E.; Stevens, J.; Commander, L.; Davis, R.; Fowler, W. EcoCheck: Perth's Banksia Woodlands Are in the Path of the Sprawling city. The Conversation 2016. Available online: https://theconversation.com/ecocheck-perths-banksia-woodlands-are-in-the-path-of-the-sprawling-city-59911 (accessed on 27 October 2018).

43. Government of Western Australia. Reimagining Perth's Lost Wetlands: Have You Ever Wondered What Perth Was Like before It Was a City? Available online: http://museum.wa.gov.au/explore/wetlands (accessed on 22 October 2018).

44. South West Aboriginal Land & Sea Council. Kaartdijin in Noongar—Noongar Knowledge: Harining Noongar Culture. Available online: https://www.noongarculture.org.au/ (accessed on 27 October 2018).

45. Jennings, G. *Tourism Research*; John Wiley & Sons: Milton, QLD, Australia, 2001.

46. Simpson, G.; Parker, J. Data for an Importance-Performance Analysis (IPA) of a Public Green Infrastructure and Urban Nature Space in Perth, Western Australia. *Data* **2018**, *3*. in press.

47. Western Suburbs Regional Organisation of Councils (WESROC). Western Suburbs Greening Plan. Available online: https://www.nedlands.wa.gov.au/sites/default/files/Western%20Suburbs%20Greening%20Plan.pdf (accessed on 5 December 2018).

48. Simpson, G.; Newsome, D.; Day, A. Data from a survey to determine visitor attitudes and knowledge about the provisioning of wild dolphins at a marine tourism destination. *Data Brief* **2016**, *9*, 940–945. [CrossRef] [PubMed]

49. Edwards, A.L. *Statistical Methods for the Behavioral Sciences*; Holt, Rinehart and Winston: New York, NY, USA, 1962; pp. 301–304.

50. Lundberg, E. The importance of tourism impacts for different local resident groups: A case study of a Swedish seaside destination. *J. Destin. Mark. Manag.* **2017**, *6*, 46–55. [CrossRef]

51. Crilley, G.; Weber, D.; Taplin, R. Predicting visitor satisfaction in parks: Comparing the value of personal benefit attainment and service levels in Kakadu National Park, Australia. *Visit. Stud.* **2012**, *15*, 217–237. [CrossRef]

52. McGuiness, V.; Rodger, K.; Pearce, J.; Newsome, D.; Eagles, P.F. Short-stop visitation in Shark Bay World Heritage Area: An importance–performance analysis. *J. Ecotour.* **2017**, *16*, 24–40. [CrossRef]

53. Patroni, J.; Newsome, D.; Kerr, D.; Chan, J.K.L.; Teo, A.C.K.; Simpson, G.D. Applying Importance-Performance Analysis to Inform Future Marine Wildlife Tourism. *J. Tour. Futures* **2018**, under review.

54. Albaum, G. The Likert scale revisited. *Mark. Res. Soc. J.* **1997**, *39*, 1–21. [CrossRef]

55. Babbie, E.R. *The Practice of Social Research*, 6th ed.; Wadsworth Publishing Company: Belmont, CA, USA, 1992; ISBN 978-1-133-59414-7.

56. Sarantakos, S. *Social Research*, 2nd ed.; Macmillan Education Australia Pty. Ltd.: South Yarra, VIC, Australia, 1998; ISBN 978-0-230-29532-2.

57. Ryan, C.; Cressford, G. Developing a visitor satisfaction monitoring methodology: Quality gaps, crowding and some results. *Curr. Issues Tour.* **2003**, *6*, 457–507. [CrossRef]

58. Lin, B.B.; Fuller, R.A.; Bush, R.; Gatson, K.J.; Shanahan, D.F. Opportunity or orientation? Who uses urban parks and why. *PLoS ONE* **2014**, *9*. [CrossRef]

59. Child, S.T.; McKenzie, T.L.; Arrendondo, E.M.; Elder, J.P.; Martinez, S.M.; Ayala, G.X. Associations between park facilities, user demographics, and physical activity levels at San Diego County parks. *J. Park Recreat. Adm.* **2014**, *32*, 68–81.

60. Matsuoka, R.H.; Kaplin, R. People needs in the urban landscape: Analysis of landscape and urban planning contributions. *Landsc. Urban Plan.* **2008**, *84*, 7–19. [CrossRef]

61. Johnson, A.J.; Glover, T.D. Understanding urban public space in a leisure context. *Leis. Sci.* **2013**, *35*, 190–197. [CrossRef]

62. Australian Bureau of Statistics, Claremont (T) (LGA) (51750), Western Australia, People & Population. Available online: http://stat.abs.gov.au/itt/r.jsp?RegionSummary&region=51750&dataset=ABS_REGIONAL_LGA&geoconcept=REGION&datasetASGS=ABS_REGIONAL_ASGS&datasetLGA=ABS_REGIONAL_LGA&regionLGA=REGION&regionASGS=REGION (accessed on 27 October 2018).

63. Australian Bureau of Statistics, Greater Perth (GCCSA) (5GPER), Western Australia, People & Population. Available online: http://stat.abs.gov.au/itt/r.jsp?RegionSummary&region=5GPER&dataset=ABS_REGIONAL_ASGS&geoconcept=REGION&datasetASGS=ABS_REGIONAL_ASGS&datasetLGA=ABS_NRP9_LGA&regionLGA=REGION&regionASGS=REGION (accessed on 27 October 2018).

64. Siu, V.W.; Lambert, W.E.; Fu, R.; Hillier, T.A.; Bosworth, M. Build environment and its influences on walking among older women: Use of standardized geographic units to define urban forms. *J. Environ. Public Health* **2012**, *2012*, 203141. [CrossRef]

65. Lindenmayer, D.; Bennett, A.; Hobbs, R. How far have we come? Perspectives on ecology, management and conservation in Australia's temperate woodlands. In *Temperate Woodland Conservation and Management*; Lindenmayer, D., Bennett, A., Hobbs, R., Eds.; CSIRO PUBLISHING: Collinwood, VIC, Australia, 2010; pp. 363–374, ISBN 978-0-643-10037-4.

66. Munro, N.; Lindenmayer, D. *Planting for Wildlife: A Practical Guide to Restoring Native Woodlands*; CSIRO PUBLISHING: Collinwood, VIC, Australia, 2012; ISBN 978-0-643-310312-2.

Article

The Usage and Perception of Pedestrian and Cycling Streets on Residents' Well-being in Kalamaria, Greece

Thomas Panagopoulos [1,*], Stilianos Tampakis [2], Paraskevi Karanikola [2], Aikaterini Karipidou-Kanari [2] and Apostolos Kantartzis [2]

[1]   Research Centre of Tourism, Sustainability and Well-Being, Faculty of Science and Technology, University of Algarve, Gambelas Campus, 8000 Faro, Portugal

[2]   Department of Forestry and Management of the Environment and Natural Resources, Democritus University of Thrace, 193 Pantazidou Street, 68200 Orestiada, Greece; stampaki@fmenr.duth.gr (S.T.); pkaranik@fmenr.duth.gr (P.K.); katerinakaripidoukanari@hotmail.com (A.K.-K.); apkantar@fmenr.duth.gr(A.K.)

*   Correspondence: tpanago@ualg.pt; Tel.: +351-289-800-900

Received: 18 July 2018; Accepted: 27 August 2018; Published: 30 August 2018

Abstract: Pedestrian zones are public spaces intended for the continued and safe mobility of pedestrians and people with disabilities, and they provide multiple benefits to urban areas. They counterbalance the densely built-up areas, decrease atmospheric pollution, increase available green or social space, increase walking and cycling rates, and facilitate active play for children. Done properly, pedestrianization may also increase local business sales. Greece boasts open public spaces and the pedestrianization of common roads. The economic crisis that Greece has been experiencing since 2008 has led people to give up their vehicles and use the pedestrian streets more frequently. The purpose of this paper was to investigate residents' perceptions and satisfaction rates concerning the pedestrian streets of Kalamaria, Greece, and evaluate their importance for residents' well-being. Following a random sampling method, 400 residents were interviewed. A two-step cluster analysis was conducted. The survey showed that the urban residents visited pedestrian zones in Kalamaria at least once a week, and the visits lasted 46–60 min. The improvement of urban landscape aesthetics and people's health and well-being were evaluated as important functions of pedestrian zones. The results also indicate that residents were not satisfied with their quality of life and the existing green infrastructures of the pedestrian streets, even though they have a positive disposition toward the construction or transformation of pedestrian streets. The residents expressed their unwillingness to pay more public taxes for the construction and maintenance of pedestrian and cycling streets. The safety and convenience of the mobility of residents were the most important advantages of the pedestrian streets. Meanwhile, overspill parking and difficulties with finding parking spaces were the main disadvantages for the residents. Local authorities can use the results of the present survey to manage the city's green infrastructure and use this information in the urban planning framework.

Keywords: pedestrian zones; well-being; viable city; residents' views; green infrastructure; Greece; biophilic urbanism

---

1. Introduction

Pedestrian zones are public areas of a city or a town reserved only for the use of pedestrians in which most or all automobile traffic is prohibited. Converting a street or a bigger area to pedestrian use only is called pedestrianization [1]. Until the end of the 19th century, the squares and central streets of cities and towns served to meet the needs of the residents. Urban public spaces were created to fulfill the everyday needs of the residents, such as communication and entertainment [2]. These places have also been used for social, civil, commercial, and political functions [3] since the middle of the 20th century, when the use of the car took a dominant role in peoples' lives [4].

The cities of today have requirements that differ from those of the past [5,6]. Nowadays, there is a universal aim for urban "regeneration" focused on urban planning that gives importance to walkability, i.e., the easiness of freely moving within the urban context—a freedom that must be ensured to increasingly wider proportions of the population [7,8]—so it is necessary to reconsider how the pedestrian moves and how the pedestrian infrastructure characteristics affect such behavior [9].

Pedestrians include all people who walk through shopping and service areas, from home to a friend's house, and take typically short trips. Every trip begins and ends as a pedestrian action, so everyone is a pedestrian at regular and various times and places in their lives [10]. Walkability is recognized as an important factor for both the quality of urban space and people's quality of life [11]. It is a spatial requisite of the built environment that greatly contributes to its livability and enables people to more effectively and fully use—and benefit from—urban opportunities. The possibility for people of different ages, genders, residential locations, socio-economic status, and personal abilities to reach valuable destinations and places "on their own" and by foot is considered an important capability offered in a sustainable city [12].

From this point of view, pedestrian infrastructure quality is very important. In fact, the lack of quality and accessibility of pedestrian areas from home or work locations leads to the exclusion of the citizens from economic, social, and cultural progress [13]. Regardless of the type of pedestrian or purpose of the trip, all pedestrians have basic needs. Safety is the primary need for pedestrians, who are often the most exposed to the dangers of vehicle traffic. They require safe access to multiple services.

Meanwhile, pedestrian zones have different effects on urban areas. Their main uses are the free mobility of pedestrians and the development of local business activities in the specific areas [14]. However, pedestrian zones are also opportunities to increase the urban green infrastructure, reintroduce nature in urban areas, and compensate for the density of buildings. Biophilic urbanism aims to bring more nature into the city and utilize green infrastructure to improve people's health and well-being [15].

There is wide interest in pedestrianization and pedestrian zones in different countries: in the United States (U.S.A.) [14], Turkey [10], India [16], and in several European countries such as Lithuania [2], the United Kingdom (U.K.) [17], and Germany [18]. The findings of these studies highlight that the pedestrian zones affect not only economic factors related to tourism development [19], job creation [20], and incentives for small and medium-sized business [21], they also have environmental impacts, including noise and atmospheric pollution reductions, and social impacts, including increasing the safety and enhancing the appearance of urban areas [22–25]. Therefore, it may be assumed that one of the ways to return quality of life to cities is to dedicate all open urban spaces to pedestrians and cyclists [2].

Besides walking, cycling is another form of active and sustainable mobility for short trips [26,27]. Although cycling was a common means of transport in Greece until the 1970s, more recently, the lack of safety standards has forced many people to use cars for their daily transportation needs. In recent years, we have witnessed an effort to reintroduce cycling in the life of people living in cities [28]. The first cycle networks in Greek cities have already been implemented, mirroring other European countries [29]. In Greece, there is confusion with regard to the open spaces where cycling is permitted. Pavements, parks, and pedestrian zones are areas that cyclists use, but pedestrians usually complain [30,31]. Previous studies in Greece have explored the intention of residents to cycle in three big cities [28–30], and some recorded the suitability for cycling in two small cities [27,32].

The aim of this study was to determine the perception of residents about the main functions of pedestrian zones that affect residents' well-being in a Greek municipality. The frequency and duration of residents' visits was also recorded, and the suitability of the municipality for cycling and the use of parks and pedestrian zones to cycling were evaluated.

This paper is organized into four sections: the literature review; information about the study area and methodology; description of the survey results; and finally, in the last section, the main conclusions are discussed and recommendations are provided for decision-makers about how to promote walking and cycling in municipalities with similar characteristics.

2. Materials and Methods

2.1. Study Area

The data used in this paper was gathered from the municipality of Kalamaria, covering 6.4 km^2 and housing 91,279 residents. It is the second largest municipality in the metropolitan area of Thessaloniki, Northern Greece. Two-thirds of Kalamaria is surrounded by sea, with 6.5 km of attractive coastline. In the post-war period, it was separated from the main city of Thessaloniki; meanwhile, the border between the two municipalities exists purely for administrative purposes, as it constitutes a residential and recreational area for Thessaloniki. The population of Kalamaria has increased rapidly by 12.4% in the last decade, which has been mainly due to the relocation of families from Thessaloniki to suburban areas. The municipality is facing severe environmental pressure due to rapid urbanization. The green areas of the municipality of Kalamaria cover 635,800 m^2, which account for 9.93% of the municipality's total area.

According to Greek Law 4315/2014 (FEK 269/24-12-2014), the pedestrian zones are open public spaces intended for the continued and safe mobility of pedestrians and people with disabilities. Cars are allowed only for residents moving to the entrance and exit of private parking spaces. The pedestrian zones in Greece comprise pedestrian streets that are frequently derived from the pedestrianization of common roads and a green infrastructure network for the unification of public use areas and social services. Figure 1 presents this green densification network, as green dots, for the promotion of walking and cycling in Kalamaria, which constitutes a good opportunity for a sustainable and biophilic urban design.

2.2. The Survey

The population under study included all of the households in the municipality of Kalamaria. The applied sampling framework involved the lists of domestic electricity consumers. The use of households is a familiar case of using teams instead of sample units. It is easier and more affordable [33]. Structured face-to-face interviews were conducted, and simple random sampling was used [34,35]. The average duration of the interview was 20 min. The survey was divided into four different sections:

1. General demographics of respondent
2. Contribution of pedestrian streets to respondent's quality of life
3. The suitability of the municipality for cycling
4. Advantages and disadvantages of pedestrian streets

Data were collected in 2014 between the months of April and June. The households were found randomly, using tables of random numbers. A personal interview was conducted for one family member per household. The response rate of the survey was very high (97.5%). Participants had to be at least 18 years old due to legal constraints in Greece. If a member of the specific household was not found or refused to complete the questionnaire, we proceeded to new sample units.

The population proportion p, as well as the estimation of the standard error of the population s$_p$ for qualitative data, questions, and the mean and standard deviation s for the quantitative data was carried out through the use of the simple random sampling formulas. To determine the sample size, pre-sampling was used, with a sample size of 50 individuals. The size of the sample was estimated according to the simple random sampling formulas where t = 1.96 and e = 5% (for qualitative data), and where t = 1.96 and e = 0.35 (for quantitative data) [36]. A total of 400 questionnaires were collected from the municipality of Kalamaria at the east of the city of Thessaloniki.

Figure 1. Masterplan of Kalamaria municipality showing the green areas and the green infrastructure network for the unification of public use areas and social services. Adapted from the revised Masterplan of Kalamaria municipality (FEK 3/AAP/15-1-2015).

Reliability analysis was applied in the multi-theme variables concerning the advantages and disadvantages of pedestrian streets. In particular, to find out the internal reliability of a questionnaire [37], we used the α coefficient (or reliability coefficient, Cronbach's α). A coefficient α that is equal to or higher than 0.70 is considered satisfactory [38], while higher than 0.80 is considered very satisfactory. In practice, the reliability coefficients with values lower than 0.60 have also been accepted many times [39]. The validity of the test was checked through factor analysis, aiming to discover the existence of common factors within a group of variables [40].

Regarding the significance of the principal components, we used the criterion suggested by Guttman and Kaiser [37]. The appropriate number of principal components was determined by the values of typical roots equal to or higher than one. Furthermore, we also used the matrix rotation of the main factors, applying Kaiser's method of maximum variance rotation. Finally, we examined the components that could explain the correlations among the variables of the data, and also attempted to provide an interpretation [41]. The variables that "belong" to each factor were those whose loadings, in the table representing the loadings of the factors after rotation, were higher than 0.5 for that factor.

A statistical segmentation of the residents in three distinct groups (clusters) was undertaken according to the advantages and disadvantages of pedestrian streets from the factor analysis (continued variables) and the acceptance of the transformation of new pedestrian streets, and how frequently they were used (categorical variables) A two-step cluster analysis was chosen for this purpose. This method constitutes a research tool that helps determine clusters with variables of the same characteristics in a large number of data (questionnaires). Considering that the variables were independent of one another, categorical and continued variables were handled at the same time following the polynomial and the normal distribution, respectively [42]. Additionally, the correlation of the other variables (continued or categorical) in every cluster separately was identified with a check of Pearson's X^2. In this way, the identity of every cluster was determined with more accuracy. The statistical package SPSS 16 was used for the data analysis [43].

3. Results

3.1. Demographic Profile of the Respondents

During the interviews, the residents were initially asked about their demographic profile. As shown in Table 1, 45.7% (s_p = 0.0254) of the respondents questioned were male, and 54.3% (s_p = 0.0254) were female. Most of them (29.9%, s_p = 0.0234) were middle-aged (31–40 years), married (57.4%, s_p = 0.0252), and without children (43.9%, s_p = 0.0253). Regarding their profession, they were mainly public servants (24.9%, s_p = 0.0221) or private employees (31.4%, s_p = 0.0237). Their educational level was quite high, since over 38.4% (s_p = 0.0248) of the respondents had completed upper secondary school or technological education (21.8%, s_p = 0.0211).

Table 1. Socio-demographic profile of the sample (s_p: Standard error of proportion).

		p (%)	s_p
Gender	Male	45.7	0.0254
	Female	54.3	0.0254
Age	18–30	20.8	0.0207
	31–40	29.9	0.0234
	41–50	26.5	0.025
	>50	20.3	0.0205
	No answer	2.6	0.0081
Marital status	Unmarried	28.8	0.0230
	Married	57.4	0.0252
	Divorced/widowed	11.7	0.0156
	No answer	2.3	0.0077
Number of children	Without children	43.	0.0253
	One child	17.1	0.0192
	Two children	30.9	0.0236
	Three children	5.2	0.0113
	More than three	2.9	0.0085
Educational level	Primary School	9.1	0.0147
	Lower Secondary	5.5	0.0116
	Upper Secondary	38.4	0.0248
	Technical School	6.8	0.0128
	Technological educ.	21.8	0.0211
	University	14.5	0.0180
	No answer	3.9	0.0099
Profession	Private employee	31.4	0.0237
	Public servant	24.9	0.0221
	Self-employed	23.6	0.0217
	Farmer	0.5	0.0037
	Pensioner	3.4	0.0092
	Student	7.8	0.0137
	Homemaker	3.6	0.0096
	Unemployed	2.9	0.0085
	No answer	1.8	0.0068
Annual income	≤5,000 €	8.3	0.0141
	5,001–10,000 €	14.3	0.0179
	10,001–20,000 €	26.0	0.0224
	20,001–30,000 €	15.3	0.0184
	>30,000 €	13.5	0.0174
	No answer	22.6	0.0213

3.2. Appraising Residents' Perception of Pedestrian Streets

There is an increasing demand for pedestrian-friendly communities to the detriment of car-oriented developments all over the world [44]. This trend is also followed by the residents of the municipality of Kalamaria, who perceive the transformation of traffic roads to pedestrian streets positively. Two-thirds of the study population 64.5% (s_p = 0.0240) were positive to the formation of new pedestrian streets; 21% (s_p = 0.0204) were either positive or negative. Only 8% (s_p = 0.0136) were negative, and 6.5% (s_p = 0.0123) did not answer the question.

The urban environment and landscape (urban scape) can cease to be synonymous with stress and compulsion. It is the right of every resident in a city to walk and observe without stress, and through this observation better know and love his or her city [7]. In the Greek language, the word "omorfo" (meaning beautiful) is something that has a good external shape. The human eye needs time to see the form (shape) of things. Walking and moving slowly, our eyes evaluate the space they are moving through much better. As shown in Figure 2, the improvement of an urban landscape was rated higher by the residents (53.3% as high and 19.8% very high) due to its contribution to pedestrian streets. The contribution to residents' psychology (48.0% as high and 18.0% very high) was also rated highly. Additionally, economic development and the chance for recreation and sports were rated lower by the residents.

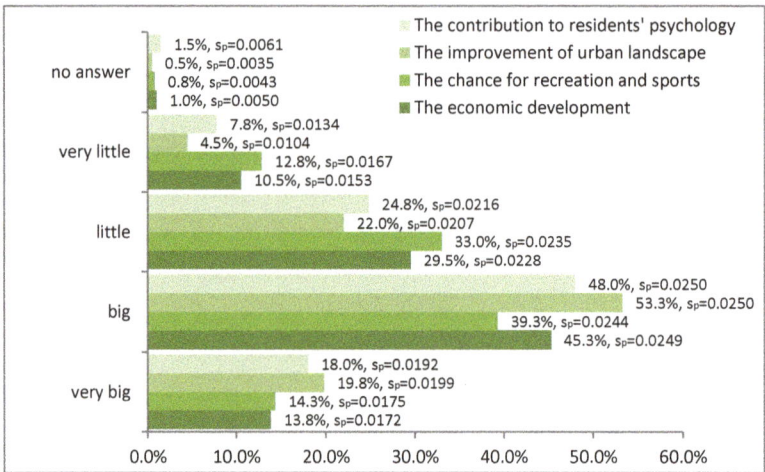

Figure 2. Contribution of pedestrian streets to residents' quality of life.

Two-thirds of the residents in the municipality perceived their own well-being as little satisfied (54.8%, s_p = 0.0254) or not at all satisfied (11.2%, s_p = 0.0161) with their quality of life in the city. Besides, the combination of the current economic downturn and the decrease of available money for recreational activities have highlighted open green spaces as cost-free alternatives [45].

Plantings in the pedestrian zones should create desirable microclimates and contribute to the psychological and visual comfort of users. Planting design and plant choices for areas surrounding pedestrian areas play a big role in the overall appearance and environmental impact of the pedestrian area installation or new development [10].

In previous research conducted by Karanikola et al. [46] in the municipality of Kalamaria, the residents were little satisfied with the existing green infrastructure. In our present research the residents have the same opinion for the existing green infrastructure in pedestrian zones. According to the results, 47.5% (s_p = 0.0250) were little and 11.5% (s_p = 0.0160) were not at all satisfied with the existing planted trees and plants; 35.5% (s_p = 0.0240) were satisfied, 0.5% (s_p = 0.0035) were absolutely satisfied; 4.5% (s_p = 0.0104) were very satisfied; 0.5% (s_p = 0.0035) did not answer the question. At this point, we must explain that the term green infrastructure includes trees that are not only in woodlands but also in streets and provide important ecosystem services in urban population [5,47]. On the contrary, residents were satisfied with the design of the existing pedestrian streets. Specifically, more than half of them were satisfied 50.5% (s_p = 0.0250), very satisfied 9.5% (s_p = 0.0147), or absolutely satisfied 1.5% (s_p = 0.0061) with the design of new pedestrian streets. In total, 31.5% (s_p = 0.0233) of the residents stated that they were less satisfied, and 6.3% (s_p = 0.0121) were not at all satisfied. In total, 0.8% (s_p = 0.0043) of the residents did not answer the question.

In the third part of the survey, the residents were asked about the frequency and the duration of their use of pedestrian streets. According to the results, the majority of residents 73% (s_p = 0.0222) visited the pedestrian streets at least once a week, 15.5% (s_p = 0.0181) a few times per month. In total, 1.3% (s_p = 0.0056) used the pedestrian streets of their municipality for walking only a few times per year or rarely 9.3% (s_p = 0.0145). In total, 1% (s_p = 0.0050) did not answer the question. Regarding the duration of their visit in the pedestrian streets, 24% (s_p = 0.0214) of the residents stated that their visits lasted less than 45 min, 31.75% (s_p = 0.0233) walked on the pedestrian streets for 46–60 min, 25.75% (s_p = 0.0219) spent less than 150 min, and 13.5% (s_p = 0.0171) spent more than 150 min. In total, 5% (s_p = 0.0109) of the respondents did not answer the question. Comparing the results of a similar study conducted in three cities of Lithuania [2], the frequency of their visits was lower; only 14% of the residents visited the pedestrian zones of their cities once or twice per week, while 44% of them visited them once or twice per month.

Pedestrianization schemes are often associated with increased retail turnover and increased property values locally [23]. The property value will be higher in a community where one can quickly and comfortably walk to and from local amenities (home to school, parks, and stores) [48]. Regarding the opinion that walkability raises property values, two-thirds 60.8% (s_p = 0.0244) of the residents in Kalamaria stated that it has a positive association, and only 4.3% (s_p = 0.0101) stated that it has a negative association. One-third 34.5% (s_p = 0.0249) of them stated that walkability neither raises nor reduces the property values, and 0.5% (s_p = 0.0035) did not answer the question. On the contrary, when they were asked to pay more public taxes for the construction and maintenance of the pedestrian zones (squares and streets), only one-third responded positively (34.3% s_p = 0.0238). Due to the current economic crisis, the majority, 64% (s_p = 0.0240) did not intend to pay more money. In total, 1.8% (s_p = 0.0066) did not answer the question.

Walking is the most affordable and accessible mode of transport, but the second most sustainable form of mobility is cycling. The global trend for urban generation focuses on an urban landscape that benefits pedestrians and cyclists and places importance on urban mobility [7]. However, in Greece, the use of the bicycle is limited, and is quite popular only in small-sized cities [25].

According to the results in the municipality of Kalamaria, about two-thirds of the residents 63.5% (s_p = 0.0241) are positive to the use of a bicycle; one-third 29% (s_p = 0.0227) were indifferent; and only 7.3% (s_p = 0.0130) were negative to the use of a bicycle. In total, 0.3% (s_p = 0.0025) did not answer the question. However, they think that their municipality is not suitable for cycling. More specifically, only 4.5% (s_p = 0.0104) of residents are of the opinion that the municipality of Kalamaria was absolutely suitable for cycling for their transportation, and 9% (s_p = 0.0143) considered it to be very suitable. The remaining respondents characterized Kalamaria's suitability for cycling as moderate 25.8% (s_p = 0.0219), little 42.3% (s_p = 0.0247), or not at all 18% (s_p = 0.0192). Comparing these results with a similar study in Preveza, a smaller Greek city, the views are completely indifferent. The vast majority, 90.5% of the residents, were positive to the use of a bicycle, and only 1% were negative [25]. The economic crisis led the residents to refuse for pay more public taxes for the construction and maintenance of a cycling route (58.5%, s_p = 0.0247 were negative and 38.8%, s_p = 0.0244 expressed a positive view). In total, 2.8% (s_p = 0.0072) did not answer the question. However, residents recognized the problems created by pedestrian sidewalks (85%, s_p = 0.0179 expressed a negative view and only 13.2%, s_p = 0.0170 were positive), and pedestrian zones (59%, s_p = 0.0249 disagree and 40%,

$s_p = 0.0245$ agree). However, the majority accepted the use of cycling in parks (54.8%, $s_p = 0.0249$ agree and 44.8%, $s_p = 0.0249$ disagree).

3.3. Advantages and Disadvantages of Pedestrian Zones

Pedestrian zones enhance accessibility and mobility for pedestrians and improve the attractiveness of the local environment [49]. In order to determine how respondents perceived the existence of pedestrian streets, different variables were examined after completing the literature review. Some of these factors were positive (advantages), and some of them were negative (disadvantages). The evaluation of the advantages using a 10-point Likert scale (1 insignificant and 10 most important) is given in Figure 3.

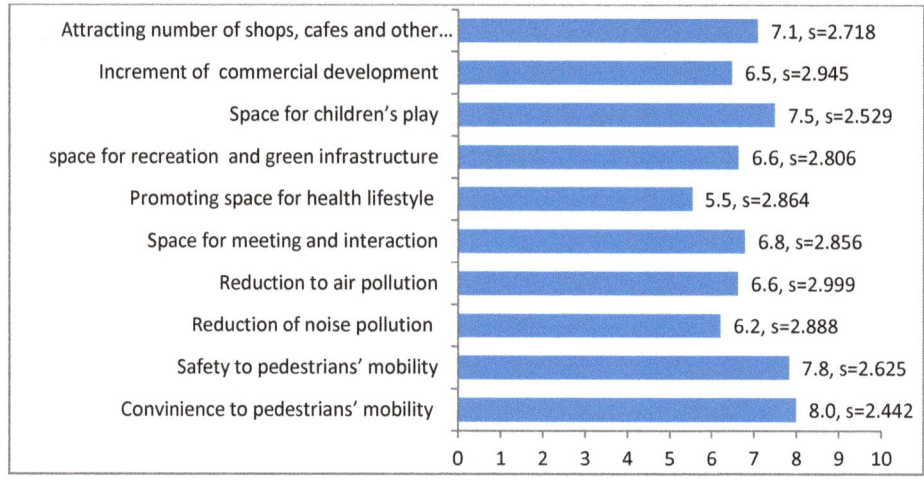

Figure 3. Evaluation of the advantages of pedestrian streets using a 10-point Likert scale (means and standard deviations).

According to this evaluation, the most important advantages were the ease of pedestrians' mobility, the safety of pedestrians' mobility, and space for children's play. Also important were the variables "Attracting a number of shops", "Space for meeting and interaction", "Reduction of atmospheric pollution", and "Green space and space for recreation". The results in a similar study conducted by Dičiūnaitė-Rauktienė et al. [2] in three Lithuanian cities were similar. The most important variable was also comfortable and safe space for pedestrians.

We applied a reliability analysis to the above variables after completing all of the necessary checks. The value of the reliability coefficient alpha is 0.894. This constitutes a strong indication that our data has the tendency to measure the same thing. In fact, this is also supported by the significantly high partial reliability coefficient alpha after the deletion of any variable, since even then, no increase in the reliability coefficient is observed. Additionally, before proceeding with the application of the factor analysis, we conducted all of the necessary checks. The value of the Keiser–Meyer–Olkin indicator is 0.841. Furthermore, Bartlett's test of sphericity rejects the null hypothesis that the correction table is unitary and that the partial correlation coefficients are low. That the measures of sampling adequacy have high to very high values also supports the view that the factor analysis model is acceptable. Two factors are extracted. Table 2 presents the loads that are the partial correlation factors of the 10 variables with each of the three factors resulting from the analysis. The higher the load of a variable in a factor, the more this factor is responsible for the total degree fluctuation of the considered variable.

Table 2. Factor burdens after rotation for the multivariable advantages of pedestrian streets.

Variable	Factor Burdens after the Rotation	
	1	2
Convenience of pedestrians' mobility	0.544	0.590
Safety of pedestrians' mobility	0.520	0.621
Reduction of noise pollution	0.206	0.862
Reduction of air pollution	0.099	0.888
Space for meeting and interaction	0.683	0.350
Promoting space for a healthy lifestyle	0.543	*0.540*
Space for recreation and green infrastructure	0.797	0.228
Space for children's play	0.790	0.042
Increment of commercial development	0.738	0.243
Attracting a number of shops, cafes, and other service vendors	0.549	0.260

The variables that "belong" to each factor are the ones for which the load (columns 1, 2) is higher than 0.5 in this factor. Factor 1: the variables "Space for meeting and interaction", "Promoting space for a healthy lifestyle", "Green space and space for recreation", "Space for children's play", "Increment of commercial development", and "Attracting a number of shops, cafes, and other service vendors)" were classed as promoting the quality of life. The second factor named "simulation of the natural environment" included the variables "Reduction of noise pollution" and "Reduction of air pollution". The variables "Ease of pedestrians' mobility" and "Safety of pedestrians' mobility" belong to the first and the second factor. Therefore, the variable "Promoting space for health and lifestyle" had a value higher than 0.5, constituting a bridge between the first and the second factor.

In Greece, sometimes the residents who claim to facilitate pedestrianizations are the ones who negate the pedestrian zones with their actions. These actions may be the parking of vehicles or the mobility of vehicles in pedestrian streets. All of the disadvantages of the existing pedestrian streets in the municipality of Kalamaria were evaluated by the residents. The results are given in Figure 4. The most important factors, as evaluated by the residents, were the parking of vehicles on pedestrian streets, difficulties in finding a parking space for their car, the mobility of vehicles in pedestrian streets, and the increase in traffic of other roads. Although pedestrian streets are used only for pedestrians and allow the entrance and exit of vehicles to private parking spaces and mobility for emergency vehicles or vehicles with supplies, in Greece, these rules are encroached because of a lack of private parking spaces in Kalamaria.

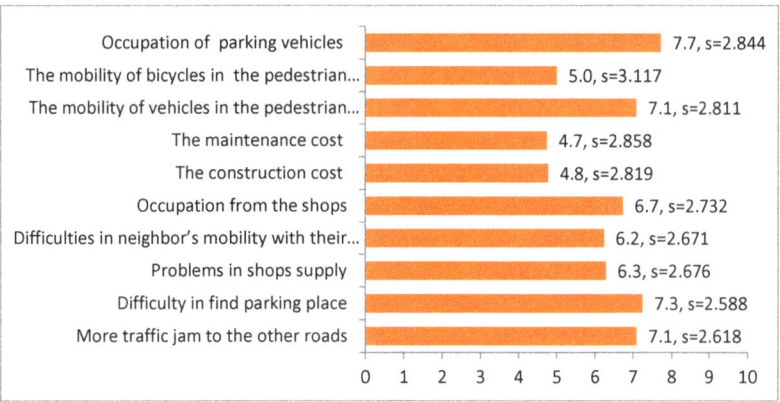

Figure 4. Evaluation of the disadvantages of pedestrian streets using a 10-point Likert scale (mean and standard deviations).

We applied reliability analysis to the above variables after completing all of the necessary checks. The value of the reliability coefficient alpha is 0.841. This constitutes a strong indication that our data has the tendency to measure the same thing. In fact, this is also supported by the significantly high partial reliability coefficient alpha after the deletion of any variable, since even then, no increase of the reliability coefficient is observed. Additionally, before proceeding with the application of the factor analysis, we conducted all of the necessary checks. The value of the Keiser–Meyer–Olkin indicator is 0.762. Furthermore, Bartlett's test of sphericity rejects the null hypothesis that the correction table is unitary and that the partial correlation coefficients are low. That the measures of sampling adequacy have high to very high values also supports the view that the factor analysis model is acceptable. Three factors are extracted. Table 3 presents the loads that are the partial correlation factors of the 10 variables, with each of the three factors resulting from the analysis.

Table 3. Table with factor burdens after rotation, for the multivariable disadvantages of the pedestrian streets.

Variable	Factor Burdens after the Rotation		
	1	2	3
More traffic on other roads	0.872	0.067	0.148
Difficulty of finding a parking space	0.893	0.138	0.147
Problems in supplying shops	0.574	0.417	0.198
Difficulties moving residents' cars	0.584	0.285	0.421
Occupation from the shops	0.224	0.570	0.193
The construction cost	0.239	0.083	0.931
The maintenance cost	0.182	0.125	0.945
The mobility of vehicles in pedestrian streets	0.113	0.840	0.084
The mobility of bicycles in pedestrian streets	0.260	0.509	0.259
Parked vehicles	0.032	0.866	−0.084

The first factor includes the variables "More traffic on other roads", "Difficulty of finding a parking space", "Problems supplying shops", and "Difficulties moving residents' cars", which we can class as "obstacles in the vehicle's mobility". We class the second factor as "occupation of their space"; this includes the variables "Occupation from the shops", "The mobility of vehicles in pedestrian streets", "The mobility of bicycles in pedestrian streets", and "Parked vehicles". The third factor is classed as "cost", and it includes the variables "construction–transformation cost" and "maintenance cost".

3.4. Correlation of Acceptance of Pedestrian Streets with Residents' Different Attitudes

The number of clusters was determined from the specific program SPSS by applying the two-step cluster analysis. The observations were grouped into three clusters as the optimum solution. More specifically, of the total sample (400 respondents), 25.9% were placed in the first cluster, 19.7% were placed in the second cluster, and 54.3% were placed in the third cluster.

Regarding the relative significance of the variables (continuous and categorical) in the formation of the clusters, the diagrammatic representations of Figure 5 present the statistical significance tests. In the case of the continuous variables, it was observed that the variable "Maintenance cost" tended to play a significant role in the formation of the first cluster, while the variables "Difficulties in vehicles' mobility" and "Occupation of pedestrian streets" were the reason for the formation of the second cluster. The variables "Improvement in quality of life" and "Simulation of nature" were the reason for the formation of the third cluster (Figure 5a,c,e).

Furthermore, regarding the categorical variables, the value of the statistical X^2 exceeded the limits of the critical value, which led to the conclusion that all of the categorical variables used in the analysis were significant for the formation of the three clusters (Figure 5b,d,f).

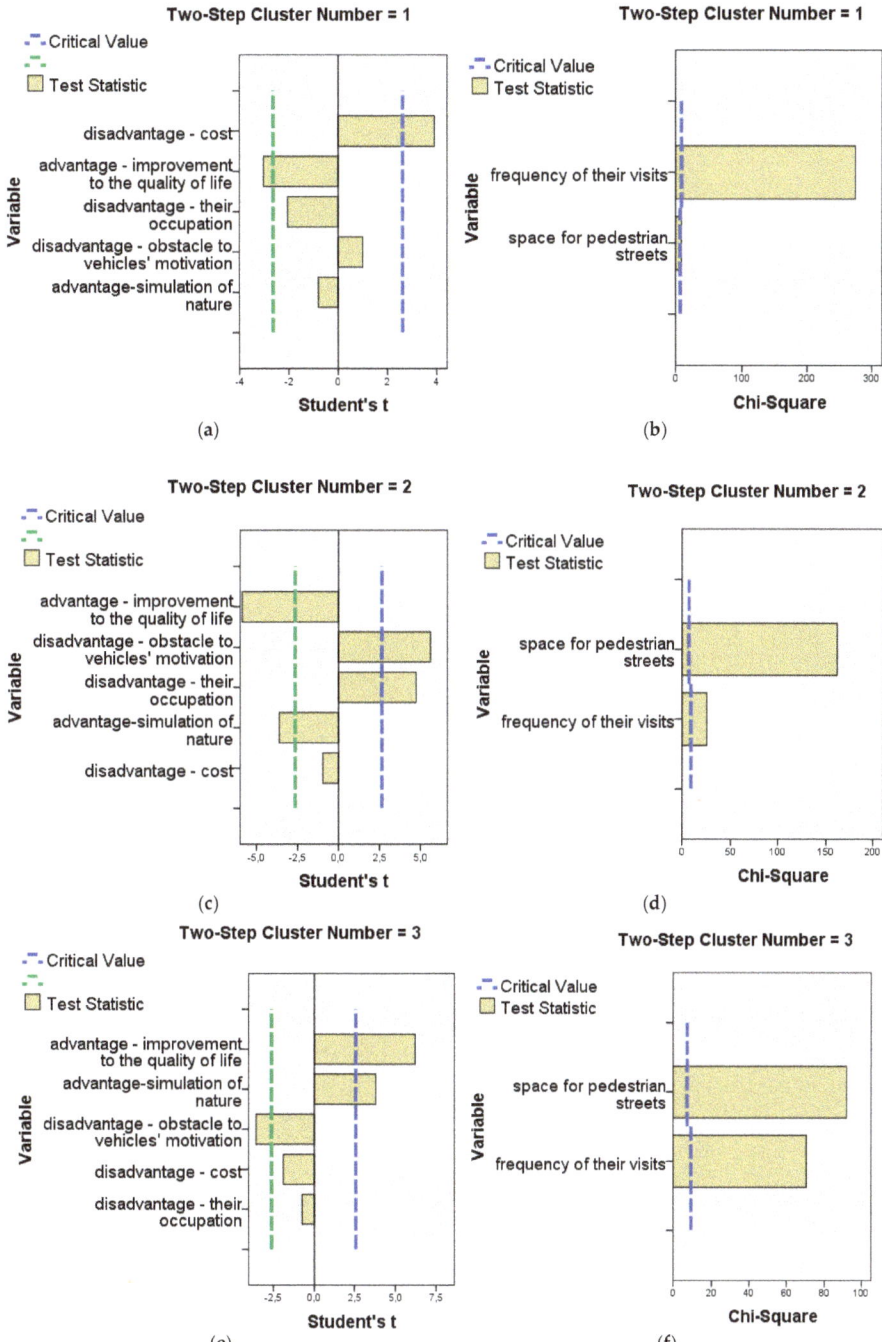

Figure 5. Diagrammatic representations of the statistical tests of variables per cluster (**a,c,e** continuous and **b,d,f** categorical variables).

From the application of the two-step cluster analysis, three clusters of residents were gathered with different characteristics among them. Table 4 presents the characteristics of the three clusters. The Pearson's X^2 test for a statistical significance of α <0.001 presented the relation of the three clusters with other quality variables.

Table 4. Interpretation of the clusters' observations.

Variables	Cluster 1	Cluster 2	Cluster 3
space for pedestrian streets	moderate or positive	negative or moderate	positive
frequency of their visits	times per month or rarely	times per week	times perweek
advantage -improvement to the quality of life	evaluated as insignificant	evaluated as insignificant	evaluated as significant
advantage-simulation of nature	evaluated as insignificant	evaluated as insignificant	evaluated as significant
disadvantage- obstacle to vehicles motivation	evaluated as limited significant	evaluated as insignificant	evaluated as insignificant
disadvantage-their occupation	evaluated as insignificant	evaluated as significant	evaluated as insignificant
disadvantage-cost	evaluated as significant	evaluated as insignificant	evaluated as insignificant
With the check of Pearson's X^2			
contribution of pedestrian streets-improvement to cityscape	significant to very significant	little significant to insignificant	significant to very significant
contribution of pedestrian street-residents' phycology	little to very little	little to very little	big to very big
contribution of pedestrian streets-economic development	little to very little	little to very little	big to very big
contribution of pedestrian streets-chance for recreation and sports	big or very big	little or very little	big or very big
duration of visit	less than 45 min.	46-150min	more than 60 min
design of pedestrian streets	little or not at all satisfied	little or not at all satisfied	absolutely satisfied or satisfied
green infrastructure of pedestrian streets	Not at all satisfied	little or not at all satisfied	very satisfied or satisfied
effect to property value	neither increase nor reduce	reduce or neither increase nor reduce	increase them
space to the use of bicycle	moderate	moderate and negative	positive
suitability of the city for cycling	very suitable, little or not at all suitable	little or not at all suitable	absolutely or very suitable
to permit bicycles in pedestrian streets	yes	no	yes
to permit bicycles in parks	yes	no	yes
more public taxes for the construction and maintenance of pedestrian streets	no	no	yes
more public taxes for the construction and maintenance of cycling net	no	no	yes

In total, 26% of the residents belong to the first cluster; they had a mediocre or positive view of pedestrian streets, and visited them for less than 45 min just a few times per month or rarely. They were little or not at all satisfied with the design of pedestrian zones and the existing green infrastructure. They were of the opinion that cost was the main disadvantage of pedestrian streets and that their property value was not affected, either positively or negatively, by the nearby pedestrian zones. They refused to pay more public taxes for the construction and maintenance of pedestrian streets and cycling routes.

In total, 19.7% of the residents belong to the second cluster; they had a negative or mediocre attitude toward pedestrian streets and visited them about once a week. The residents of this cluster associated the pedestrian streets with the disadvantages to vehicle mobility, including a lack of parking spaces. Their opinion was that pedestrian streets made little or very little contribution to either the improvement of urban landscape, residents' health, economic development, and chances for recreation and sports. Their visits to pedestrian zones last 46–150 min, and they claim to be little or not at all satisfied with the design of new pedestrian streets and the related green infrastructure. They were of the opinion that the pedestrian zones in their municipality neither increased nor reduced the property values, and the city was little or not at all suitable for cycling. The residents of that second cluster also did not accept cycling in pedestrian zones and parks, and they refused to pay more taxes for the construction and maintenance of pedestrian zones and cycling routes.

In total, 54.3% of the residents belong to the third cluster; they had a positive attitude toward pedestrian streets in their municipality and correlated them with a high quality of life and natural simulation. The residents of this cluster visited those zones at least once a week and for longer periods. They were of the opinion that pedestrian zones highly contributed to the improvement of urban landscape, human health, economic development, and increased chances for recreation and sport activities. They claimed to be completely satisfied with the design of the pedestrian streets and related green infrastructure. Additionally, they were of the opinion that the existence of the pedestrian zones increased the value of their property in the municipality. They had a positive view on the use of a bicycle, and they accepted cycling in pedestrian zones and parks. Additionally, they considered that their city was suitable for cycling, and they accepted paying more public taxes for the construction and maintenance of pedestrian and cycling zones.

4. Discussion and Conclusions

The economic crisis that Greece has been experiencing since 2008, the increasing ticket prices of public transport services, high car maintenance costs, and increased environmental consciousness have led to people using their vehicles less and less, and subsequently increasing their preference for other modes of transport (bicycles and walking) [50]. To facilitate the mobility of the residents, it is necessary to improve the existing infrastructures of the pedestrian streets and cycling routes. Low-budget strategies, as a deliberate means of creating valuable, attractive, well-used, sociable, public spaces, are sustainable solutions [45]. A number of unused and neglected spaces of often obscure property status, identity, and function can be added to those green public spaces. These urban "cracks", as Loukaitou-Sideris [51] called them, can act as informal open spaces. Despite their current condition, some urban "cracks" may have a crucial location within the urban fabric, and with low-cost interventions from the local municipality, may be linked to the existing network of open spaces.

Even though Kalamaria is fully urbanized, some natural vegetation can be found mostly on steep slopes along the coastline, which is not fully connected with the green infrastructure network. Moreover, Kalamaria lacks a citywide park, and for this reason, the statutory General Urban Plan considered the reuse of all large empty spaces such as the derelict Kodra military camp and a large part of the Ntalipi military camp [52]. The connection of the existing pedestrian and cycling zones with the coastline pedestrian zone of Kalamaria and Thessaloniki and also with other existing public green areas will provide better quality of life to residents and constitutes a good opportunity for the sustainable and biophilic design of Thessaloniki as a whole.

For this reason, the majority of Kalamaria residents (64.5%) expressed positive views about the construction or transformation of new pedestrian streets in their municipality. Additionally, they were not satisfied with the design quality of the existing infrastructure. The survey also showed that the urban residents visited pedestrian zones at least once a week, and the visit lasted 46–60 min, which is much more time compared to the results of a similar study in Lithuania [2].

Urban landscape improvement and residents' health were considered as the most important functions of pedestrian zones to residents' well-being. On the contrary, the economic development of the local market and the chance for recreation and sports were evaluated as less important. This was also influenced by improvements in technology and tools that enable online shopping, which reduced the need for shopping in brick-and-mortar stores. Moreover, a new leisure culture created in shopping centers has resulted in the public space crisis in countries such as Greece, Portugal, and Lithuania [53,54].

The results of this study revealed that the existence of pedestrian zones may contribute to increased property values. Furthermore, the residents were unwilling to pay more public taxes for the pedestrian streets, and they were unwilling to pay for the construction and maintenance of a cycle route. The general perception of residents was that their municipality was not suitable for cycling. However, they viewed the use of a bicycle positively, and they were not disturbed by the existence of bicycles in parks and pedestrian zones, corroborating other similar studies in smaller towns of Greece [26].

Although there has been a significant policy shift in which local governments are taking up increasing responsibility to ensure a safe pedestrian environment, much remains to be implemented. Representations of safer and convenient city pedestrian streets may encourage more people to walk for shorter trips, which will certainly lead to a healthier and more pleasant city [54]. Confirming the above study, the residents of Kalamaria rated the convenience and safety of pedestrians' mobility as the main advantages of pedestrian streets.

Meanwhile, the pedestrians' unobstructed mobility may lead to the limitations of other activities [55] that will create a negative opinion about pedestrianization in a significant part of the community. Consequently, urban planners should make the pedestrian streets more accessible to residents, and should also be concerned with the improvement of infrastructure facilities for car parking and means of mass transportation and cycling.

The two-step cluster analysis revealed three clusters of residents with distinct characteristics. The majority of residents (54.3%) belonged to the cluster that had a positive view about pedestrian streets and cycling, agreed to pay more taxes for green infrastructure, and usually had frequent and long visits to pedestrian zones. The smallest cluster, with 19.7% of the residents, was characterized by frequent short visits to pedestrian areas, and these residents associated pedestrian streets with vehicle mobility problems. They had a negative opinion about the new and existing design of pedestrian streets and green infrastructure. They did not accept the idea of paying more taxes for the maintenance of pedestrian zones and cycling routes, and they were generally negative to the use of bicycles in their municipality. Finally, the cluster with 26% of the residents that rarely used pedestrian streets and had a moderate view of them, stating that the benefits derived from pedestrian streets were lower than the construction cost, and that they saw no value in improving the accessibility for all to an urban green infrastructure network.

Information and training programs will be essential to the community in striving for safe walking conditions. Local authorities can use the results of the present survey to manage the city's green infrastructure and meet the needs of residents for more biophilic urbanism. Public open spaces such as pedestrian streets and other green spaces are key built environment elements within neighborhoods that encourage a variety of physical activity behaviors [56] and offer multiple benefits for human well-being [5]. Meanwhile, urban policy has failed to provide specific design guidance for the health and well-being of all of the residents [57]. This public perception survey enabled urban planners to identify preferred green infrastructure alternatives and use this information in the urban planning framework.

Analysis of the results leaves room for future work that will explore the effects of green infrastructure, such as the effects of pedestrian and cycling streets on local business, and study how to assess the feasibility of the chosen solutions and convert the results into effective policy decisions in the city governance. Based on the research results, the authors believe that urban planning should provide more inclusive green spaces that respond to the varying needs of people across all life-course stages. In future research, it will be essential to study the attitudes of vulnerable and excluded groups such as migrants, the elderly, or people with disabilities, and consider the problems that they experience in relation to accessibility in pedestrian zones.

Author Contributions: All authors contributed equally to this work.

Acknowledgments: This paper was financed by the FCT-Foundation for Science and Technology through project PTDC/GES-URB/31928/2017 "Improving life in a changing urban environment through Biophilic Design".

Conflicts of Interest: The founding sponsors had no role in the design of the study; in the collection, analyses, or interpretation of data; in the writing of the manuscript, and in the decision to publish the results.

References

1. Collins Dictionary. 2017. Available online: https://www.collinsdictionary.com/ (accessed on 31 July 2017).
2. Dičiūnaitė-Rauktienė, R.; Gurskienė, V.; Burinskienė, M.; Maliene, V. The usage and perception of pedestrian zones in Lithuanian cities: Multiple criteria and comparative analysis. *Sustainability* **2018**, *10*, 818. [CrossRef]
3. Mehta, V. Evaluating public space. *J. Urban Des.* **2014**, *19*, 53–88. [CrossRef]
4. MacLachlan, A.; Biggs, E.; Roberts, G.; Boruff, B. Urban growth dynamics in Perth, Western Australia: Using applied remote sensing for sustainable future planning. *Land* **2017**, *6*, 9. [CrossRef]
5. Panagopoulos, T.; Gonzalez Duque, J.A.; Bostenaru Dan, M. Urban planning with respect to environmental quality and human well-being. *Environ. Pollut.* **2016**, *208*, 137–144. [CrossRef] [PubMed]
6. Kopecká, M.; Szatmári, D.; Rosina, K. Analysis of Urban Green Spaces Based on Sentinel-2A: Case Studies from Slovakia. *Land* **2017**, *6*, 25. [CrossRef]
7. Knoflacher, H.; Rod, P.; Tiwary, G. How roads kill cities. In *The Endless City*; Burdett, R., Sudjic, D., Eds.; Phaidon: London, UK, 2010.
8. Delso, J.; Martin, B.; Ortega, E.; Otero, I. A model for assessing pedestrian corridors. Application to Victoria-Gasteiz City (Spain). *Sustainability* **2017**, *9*, 434. [CrossRef]
9. Willis, A.; Gjersoe, N.; Havard, C.; Kerridge, J.; Kukla, R. Human movement behavior in urban spaces: Implications for the design and modeling of effective pedestrian environments. *Environ. Plan. B Plan. Des.* **2004**, *31*, 805–818. [CrossRef]
10. Sisman, E.E. Pedestrian zones. *Adv. Lands. Archit.* **2013**, *16*, 401–426.
11. Talen, E. Pedestrian access as a measure of urban quality. *Plan. Pract. Res.* **2002**, *17*, 257–278. [CrossRef]
12. Blecic, I.; Canu, D.; Cecchini, A.; Congiut, T.; Fancello, G. Walkability and street intersections in rural-urban fringes: A decision aiding evaluation procedure. *Sustainability* **2017**, *9*, 883. [CrossRef]
13. Pinna, F.; Murrau, R. Isolated and single pedestrians and pedestrian groups on sidewalks. *Infrastructures* **2017**, *2*, 21. [CrossRef]
14. Robertson, K.A. Pedestrianization strategies for downtown planners: Skywalks versus pedestrian malls. *J. Am. Plan. Assoc.* **1993**, *59*, 361–370. [CrossRef]
15. Beatley, T.; Newman, P. Biophilic cities are sustainable, resilient cities. *Sustainability* **2013**, *5*, 3328–3345. [CrossRef]
16. Soni, N.; Soni, N. Benefits of pedestrianization and warrants to pedestrianize an area. *Land. Use Policy* **2016**, *57*, 139–150. [CrossRef]
17. Ward, S.V. What did the Germans ever do for us? A century of British learning about and imagining modern town planning. *Plan. Perspect.* **2010**, *25*, 117–140. [CrossRef]
18. Hass Klau, C. Impact of pedestrianization and traffic calming on retailing a review of the evidence from Germany and the UK. *Transp. Policy* **1993**, *1*, 21–31. [CrossRef]
19. Newman, L. The virtuous cycle: Incremental changes and a process-based sustainable development. *Sustain. Dev.* **2007**, *15*, 267–274. [CrossRef]
20. Blaga, O.E. Pedestrian ones as important urban strategies in redeveloping the community-Case study: Alba Iulia Borough Park. *Transylv. Rev. Admin. Sci.* **2013**, *38*, 5–22.
21. Carmona, M. London's local high streets: The problems, potential and complexities of mixed street corridors. *Prog. Plan.* **2015**, *100*, 1–84. [CrossRef]

22. Wicramasinghe, V.; Dissanayake, S. Evaluation of pedestrians' sidewalk behavior in developing countries. *Transp. Res. Procedia* **2017**, *25*, 4068–4078. [CrossRef]

23. Jakovlevas-Mateckis, K. Some aspects of the formation of pedestrian streets and zones in the new public spaces of urban centre. *J. Arch. Urban* **2012**, *36*, 252–263.

24. Vlachokostas, C.; Nastis, S.; Achillas, C.; Kalogeropoulos, K.; Karmiris, I.; Moussiopoulos, N.; Chourdakis, E.; Banias, G.; Limperi, N. Economic damages of ozone air pollution to crops using combined air quality and GIS modelling. *Atmos. Environ.* **2010**, *44*, 3352–3361. [CrossRef]

25. Castillo-Manzano, J.I.; Lopez-Valpuesta, L.; Asencio-Flores, J.P. Extending pedestrianization processes outside the old city center; conflict and benefits in the case of the city of Seville. *Habitat Int.* **2014**, *44*, 194–201. [CrossRef]

26. Pooley, C.; Horton, D.; Schelderman, G.; Mullen, C.; Jones, T.; Tight, M.; Jopson, A.; Chisholm, A. Policies for promoting walking and cycling in England: A view from the street. *Transp. Policy* **2013**, *27*, 66–72. [CrossRef]

27. Karanikola, P.; Panagopoulos, T.; Tampakis, S.; Tsantopoulos, G. Cycling as a smart and green mode of transport in small touristic cities. *Sustainability* **2018**, *10*, 268. [CrossRef]

28. Winters, M.; Davidson, G.; Pearce, C.; Teschke, K. Cycling in cities-Understanding people, neighborhoods and infrastructure to guide urban design for active transportation. In Proceedings of the 45th International Making Cities Livable Conference, Portland, OR, USA, 10–14 June 2007.

29. Vlastos, T.; Milakis, D. Planning of a cycling network in a Greek city according to geometrical criteria. The case of Moschato. *Tech. Chron. Sci. J.* **2003**, *23*, 35–46.

30. Milakis, D. Will Greeks cycle? Exploring intention and attitudes in the case of the new bicycle network of Patras. *Int. J. Sustain. Transp.* **2015**, *9*, 321–334. [CrossRef]

31. Papavasileiou, C.; Milakis, D.; Vlastos, T. Car dependence or appetence? Examination of attitudes towards sustainable mobility in the Greek case. In Proceedings of the 12th World Conference on Transport Research, Lisbon, Portugal, 11–15 July 2010.

32. Tampakis, S.; Karanikola, P.; Tsantopoulos, G.; Andrea, V.; Antipa, N.M.; Paroni, D.V. Exploring the positive and negative impacts of bicycling in the city of Orestiada, Greece. In Proceedings of the Virtual International Conference on Advanced Research in Scientific Areas (ARSA-2013), Bratislava, Slovakia, 2–6 December 2013; pp. 346–350.

33. Matis, K. *Forest Sampling*; Democritus University of Thrace: Xanthi, Greece, 2001.

34. Hoyos, D. The state of the art of environmental valuation with discrete choice experiments. *Ecol. Econ.* **2010**, *69*, 1595–1603. [CrossRef]

35. Pagano, M.; Gauvreau, K. *Elements of Biostatistics*; Ellin Publications: Athens, Greece, 2000.

36. Karlis, D. *Multivariate Statistical Analysis*; Stamoulis Publications: Athens, Greece, 2005.

37. Frangos, C.K. *Methodology of Market Research and Data Analysis with the Use of the Statistical Package SPSS for Windows*; Interbooks Publications: Athens, Greece, 2004.

38. Howitt, D.; Gramer, D. *Statistics with the SPSS 11 for Windows*; Kleidarithmos Publications: Athens, Greece, 2003.

39. Siardos, G.K. *Multivariate Statistical Analysis Methods. Part I: Exploring the Relations between Variables*; Zitis Publications: Thessaloniki, Greece, 1999.

40. Sharma, S. *Applied Multivariate Techniques*; John Wiley & Sons: New York, NY, USA, 1996.

41. Djoufras, I.; Karlis, D. *Elements of Multivariate Data Analysis*; University of the Aegean: Chios, Greece, 2001.

42. Harman, H.H. *Modern Factor Analysis*; The University of Chicago Press: Chicago, IL, USA, 1976.

43. *SPSS Categories 16*; A Software Package, Version 16.0; SPSS Inc.: Chicago, IL, USA, 2008.

44. Speck, J. *Walkable City: How Downtown Can Save America, One Step at a Time*; Farrar, Straus and Giroux: New York, NY, USA, 2012.

45. Herman, K.; Sbarcea, M.; Panagopoulos, T. Creating green sustainability through low-budget and upcycling strategies. *Sustainability* **2018**, *10*, 1857. [CrossRef]

46. Karanikola, P.; Panagopoulos, T.; Tampakis, S.; Karipidou-Kanari, A. A perceptual study users' expectations of urban green infrastructure in Kalamaia, municipality of Greece. *Manag. Environ. Qual.* **2016**, *27*, 568–584. [CrossRef]

47. Obrien, L.; DeVreese, R.; Atmis, E.; Olafsson, A.S.; Sievanen, T.; Brennan, M.; Sanchez, M.; Panagopoulos, T.; DeVries, S.; Kern, M.; et al. Social and environmental justice: Diversity in access to and benefits from urban green infrastructure- examples from Europe. In *The Urban Forest*; Pearlmutter, D., Calfapietra, C., Samson, R., O'Brien, L., Ostoić, S.K., Sanesi, G., del Amo, R.A., Eds.; Springer: Berlin, Germany, 2017. [CrossRef]

48. Cohen, S. Does Walkability Raise Property Values? 2010. Available online: http://www.houselogic.com/home-advice/green-living/does-walkability-raise-property-values/# (accessed on 12 May 2018).

49. Chiquetto, S. The environmental impacts from the implementation of a pedestrianization scheme. *Transp. Res. Part D Transp. Environ.* **1997**, *2*, 133–146. [CrossRef]

50. Efthymiou, D.; Antoniou, C. Understanding the effects of economic crisis on public transport users' satisfaction and demand. *Transp. Policy* **2017**, *53*, 89–97. [CrossRef]

51. Vartholomaios, A.; Papadopoulou, M.; Lafazani, P.; Paraschakis, I.; Arvanitis, A.; Sarafidis, D. Identifying 'crisis-proof' places. An assessment of public space accessibility using Space Syntax and GIS in the Municipality of Kalamaria, Greece. In Proceedings of the 10th International Congress of the Hellenic Geographical Society, Thessaloniki, Greece, 22–24 October 2014.

52. Loukaitou-Sideris, A. Cracks in the city: Addressing the constraints snd potrentisl of urban design. *J. Urban Des.* **1996**, *1*, 91–106. [CrossRef]

53. Rudokas, K. The shift of the public space paradigm in prosoviet Lithuania. *Logos* **2013**, *77*, 211–222.

54. Rahaman, K.R.; Lourenço, J.; Viegas, J.M. Perceptions of pedestrians and shopkeepers in European medium-sized cities: Study of Guimarães, Portugal. *J. Urban Plan. Dev.* **2012**, *138*. [CrossRef]

55. Ehrenfeucht, R.; Loukaitou-Sideris, E. Constructing the sidewalks: Municipal government and the production of public space in Los Angeles, California, 1880–1920. *J. Hist. Geogr.* **2007**, *33*, 104–124. [CrossRef]

56. Koohsari, M.J.; Mavoa, S.; Villanueva, K.; Sugiyama, T.; Badland, H.; Kaczynski, A.T.; Owen, N.; Giles-Corti, B. Public open space, physical activity, urban design and public health: Concepts, methods and research agenda. *Health Place* **2015**, *33*, 75–82. [CrossRef] [PubMed]

57. Douglas, O.; Lennon, M.; Scott, M. Green space benefits for health and well-being: A life-course approach for urban planning, design and management. *Cities* **2017**, *66*, 53–62. [CrossRef]

MDPI

St. Alban-Anlage 66

4052 Basel

Switzerland

Tel. +41 61 683 77 34

Fax +41 61 302 89 18

www.mdpi.com

Land Editorial Office

E-mail: land@mdpi.com

www.mdpi.com/journal/land

www.ingramcontent.com/pod-product-compliance
Lightning Source LLC
Chambersburg PA
CBHW041139120626
46547CB00020B/3040